Commercial Space Technologies and Applications

T0314352

Commercial Space Technologies and Applications

Communication, Remote Sensing, GPS, and Meteorological Satellites

Second Edition

Mohammad Razani

CRC Press
Taylor & Francis Group
Boca Raton London New York

CRC Press is an imprint of the
Taylor & Francis Group, an **informa** business

CRC Press
Taylor & Francis Group
6000 Broken Sound Parkway NW, Suite 300
Boca Raton, FL 33487-2742

First issued in paperback 2020

ISBN 13: 978-0-367-65633-1 (pbk)
ISBN 13: 978-1-138-09785-8 (hbk)

Library of Congress Cataloging-in-Publication Data

Names: Razani, Mohammad, author.
Title: Commercial space technologies and applications : communication, remote sensing, GPS, and meteorological satellites / Mohammad Razani.
Other titles: Information, communication, and space technology
Description: Second edition. | Boca Raton : Taylor & Francis, a CRC title, part of the Taylor & Francis imprint, a member of the Taylor & Francis Group, the academic division of T&F Informa, plc, 2019. | Revised edition of Information, communication, and space technology. | Includes bibliographical references and index.
Identifiers: LCCN 2018026595 | ISBN 9781138097858 (hardback : acid-free paper) | ISBN 9780429454585 (ebook)
Subjects: LCSH Outer space—Civilian use. | Space industrialization. | Artificial satellites in telecommunication. | Scientific satellites.
Classification: LCC TL794.7 .R39 2018 | DDC 629.4—dc23
LC record available at https://lccn.loc.gov/2018026595

Visit the Taylor & Francis Website at
www.taylorandfrancis.com

and the CRC Press Website at
www.crcpress.com

This book is dedicated to the memory of my parents and my sister,

my family back home, and my wife, Zahra, and our children

Maryam and Amir, whom we love and we are so proud of.

Contents

Preface

This manuscript updates the book which was written 5 years ago and in addition to adding questions and problems at the end of each chapter to make it a suitable textbook for classroom environments, it also covers the new satellites in every area of interest and the new and emerging technologies since then, which I must admit have been tremendous. Scientists and engineers are working tirelessly to provide a better quality of satellite services, to bring about higher resolution satellite imagery, and to lower the cost of satellite services so even the remotest areas on earth can benefit from its services.

Space technology is the technology that aims at conquering new frontiers in outer space and reaching the furthest skies. In the space within our reach, which is the focus of this updated book, man-made satellites are entering a new era of becoming smaller, smarter, cheaper, easier to lunch, yet providing a better quality of service that is less susceptible to interference by employing the most sophisticated new technologies.

With the rapid pace of advances in electronics, semiconductors, modulation, access, and multiplexing techniques; better performing amplifiers; antenna; rocket fuel; launchers; and so many other technologies affecting the satellite industry, five years is a long time to update new technologies. During this period, enormous research activities have taken place in the space technologies and its related fields. The author tries his best to do justice in covering the most outstanding results of these challenging yet rewarding scientific and engineering findings.

Commercial space technologies and applications have always fascinated the author, so much so that most of his time during the last four decades has involved working on the different aspects of the space technology. How space was explored by humans, how it has expanded our understanding of the universe, and how it has impacted our lives in so many different ways so far and how it will impact the lives of the generations to come, are questions that volumes of books are required to provide a comprehensive understanding of space technologies and its applications. The author has tried to provide a good understanding of the history of space and space exploration and space-related activities worldwide while explaining its various applications within three major categories: application satellites, scientific satellites, and communication satellites.

Space has become a new platform for humans to carry out research work that is not possible to do on earth. Other than discussing the existing applications of space technology, this revisited book introduces what is on the horizon that perhaps the next generation will benefit from and are now mere dreams or part of our imagination. Equally distributing energy worldwide through the use of satellites, which collect solar energy from the solar-rich areas on Earth and transmit it to different locations on earth, that are solar-poor areas by means of microwave beams, or launching spacecraft into space without rockets, are among such innovative ideas that are introduced to the readers.

Searching through the existing literature, although there is scattered information about the space technologies, there was no evidence of a single book that covers such a wide spectrum of interrelated concepts and applications. This gave the author the incentive to write this updated book.

This book is intended for college students, undergraduates as well as graduate students, professionals in space-related industries, engineers and technologists in telecommunication organizations worldwide, educators at various levels of lower and higher educational establishments, and anyone who is eager and interested in space science and technologies and its applications.

This book will introduce the space technologies that are benefiting humans in their everyday lives. As it is explained in detail in this book, space provides a unique environment to carry out scientific research that otherwise would be impossible to carry out on earth. The results of such scientific research are monumental and the effect of which could be seen in almost every aspect of our lives. Within the research laboratory facilities in the International Space Station, for example, scientists and engineers have been carrying out research in a microgravity environment during the last 16 years to improve the quality of life on earth and to prolong the longevity of human lives. The fields of medicine, pharmaceuticals, metallurgy, physics, engineering, biology, and so many other fields are benefiting from such unique research activities.

Humans are a curious species and are always seeking new frontiers to expand their horizon that can someday be conquered. This book will address such motives for outer space activities which are purely scientific and will provide a better understanding of the universe we live in and appreciate the blessings we have here on earth. The planet to which there are yet no others that could come even close to its unique environmental conditions that allows life to go on without extreme conditions.

This book will also make the readers aware of the fact that it is our responsibility to preserve this uniqueness for future generations by searching for new clean energy and a better use of our resources.

Acknowledgments

All gratitude is due to the Almighty who has given me life and has provided me with all it took to make it to this level.

I must admit that writing this book and its second edition has been a great challenge for me during which I had the responsibilities of teaching as well as Chairing the Electrical and Telecommunications Engineering Technology (ETET) Department at the New York City College of Technology of the City University of New York since 2006. The ETET Department offers ABET accredited associate and bachelor degree programs in electrical engineering technology and telecommunications engineering technology and is privileged to have very high-caliber and internationally recognized faculty who tirelessly educate the diverse population of the eager and talented students who attend the school in Brooklyn, New York, and continue their contributions to the world of science and technology.

It would have been impossible to write this book without the support of many people.

My sincere thanks goes to Nora Konopka, Publisher of Engineering books at Taylor & Francis, and her assistant Kyra Lindhorm, and Alexandra Andrejevich, the Associate Project Manager who patiently worked with me during the entire journey of writing this second edition and other members of the production team.

I would like to dedicate this book to my parents who provided me with the opportunity to obtain my higher education in a faraway country and supported me in every possible way and in all aspects of my life during their fruitful and honorable lives. I would also like to thank my sister and brothers who have always been supportive of me throughout my life. Last but not least, I would like to dedicate this writing to my wife and my children for their love, support, and patience.

Author

Mohammad Razani has extensive experience in various fields of technologies spanning from Satellite Communications to Microwave Remote Sensing and Information Technology. Having received BS-EE and MS-EE from Kansas State University and PhD in electrical engineering from the University of Kansas and having been involved in NASA-funded research at the Space Research Center of the University of Kansas for many years, he has utilized his education and research experiences in various related technologies within the last four decades. Such experience includes years of teaching and research positions in New York City College of Technology of the City University of New York, where he is now a full professor and chairman of the Electrical and Telecommunications Engineering Technology Department, a position he has held since 2007. Dr. Razani has closely worked with the International Telecommunication Union, a specialized agency to the United Nations, participating as chairman and vice chairman of several CCIR and CCITT Study Groups for more than a decade. Dr. Razani has published numerous papers in top peer-reviewed journals, presented at several domestic, regional, and international conferences, and has written several books, including *Fundamentals of Satellite Communications* in 1991 and *Information, Communication and Space Technology* in 2012 by CRC Press (http://www.crcpress.com/product/isbn/9781439841631).

More information about Dr. Razani's academic background can be found at http://www.citytech.cuny.edu/faculty/MRazani.

1

Overview

This book will provide a clear understanding of space, space technologies, space applications, and its impact on the lives of humans now and for the generations to come.

The many applications of space technologies and their impact on our lives will be explored with real life and tangible examples. The future outlook of robotics, communications and navigation, human health, and nanotechnology are presented.

Space technologies have progressed at a very rapid and astonishing pace and have affected our lives in day-to-day functions, whether at home, at work, or when we travel. We all have experienced the radio signal degradations in our cars when we travel and get away from the radio transmitters. With digital satellite radio broadcasting, we no longer have these worries, and we can listen to our favorite channel traveling across continents with CD quality. Millions every year are also switching to satellite TV with hundreds of channel capabilities. Remote and isolated areas in the developing as well as developed countries can now watch the same TV channels that people who live in the metropolitan areas watch. They can use telecommunication means, e-learning, telemedicine, GPS, and numerous other services that seemed impossible only a few short years ago. Through space technologies one can penetrate deep into the Earth, ocean, or dense jungles and extract information from thousands of miles away in space. Floods could be prevented by forecasting the amount of water resulting from snowmelts, months before it even happens. Soils of different moistures can be used for different crops to improve quality and productivity through the ability to estimate soil moisture contents from space even under vegetation cover. Oil spills can be best controlled by real-time image collection and onboard processing by using Earth resource satellite. Having worked under National Aeronautics and Space Administration (NASA) contracts for many years in these areas and having taught and carried out research projects in space-related technologies, the author has a good understanding of how space and its technologies can benefit societies through a systematic and wellplanned set of policy-making decisions.

In this book, space and its applications will be discussed and the role of satellites in communications, Earth resource utilization, weather forecasting, and other areas will be explained. The UN Commission on Sustainable

Development on its 16th Session held May 5–16, 2008, in New York* in its "Space for Sustainable Development Report" explained that space technology and its applications, such as Earth observation systems, meteorological satellites, satellite communications, and satellite navigation and positioning systems, strongly support the implementation of actions called for at the World Summit on Sustainable Development. It further emphasized that space applications are effective tools for monitoring and conducting assessments of the environment, managing the use of natural resources, providing early warnings of and managing natural disasters, providing education and health services in rural and remote areas, and connecting people around the world. In the concluding remarks of the report on the "Contribution of the Committee on the Peaceful Uses of Outer Space" to the work of the "Commission on Sustainable Development for the Thematic Cluster 2008–2009," one reads: "Space science and technology and their applications, coupled with advances made in other fields of science and technology, offer a wide range of specific tools and solutions and can enable and support States in overcoming obstacles to sustainable development."[†]

These findings and other similar evidences on the effective role of space technology and its applications in various fields have inspired the author to address these issues in this book. The author intends to expose the young generation to these findings in order to guide them through their fields of studies and educate them to adapt themselves to the needs of our generation and generations to come in order to achieve the most efficient ways of utilizing these technologies for peaceful purposes and in improving the quality of life globally.

This second edition of the book is intended as a text book for college students in engineering and engineering technology fields. Although it introduces some concepts utilizing mathematical approaches, students with calculus knowledge should have no difficulty grasping these concepts and working on the questions and exercises at the end of the chapters.

---

* http://sustaianabledevelopment.un.org/intergovernmental/csd16.
† United Nations Department of Economic and Social Affairs, Commission on Sustainable Development, 16th Session, May 5–16, 2008, New York, p. 11.

2

Commercial Space Technologies

2.1 Introduction

Space has been defined in numerous ways depending on who defines it and for what purpose. One could define *space* as a "boundless, three dimensional extents in which objects and events occur and have relative position and direction."[*]

This chapter is divided into two sections, one discussing outer space and the other the space within reach. Outer space is any location outside Earth's atmosphere, and that is how this chapter is organized. Human achievements in both frontiers are admirable and have made remarkable advances in both fronts.

Until the late 17th century, philosophers were resigned to the idea that the universe was driven by supernatural forces with little relationship between the behavior of objects on Earth and those above. This view began to change around 1609, when Johannes Kepler finally swept aside all notions of heavenly spheres and celestial clockwork, replacing them with laws of planetary motion that could accurately describe the orbits of the planets. However, it was the English scientist Isaac Newton (1642–1727) who finally showed how the movement of the planets, and of objects on Earth, could all be explained through three simple laws of motion and the effects of a force he called *gravity* that was produced by any object with a substantial mass. Newton's three Laws of Motion are as follows:

- Newton's First Law of Motion states that in order for the motion of an object to change, a force must act upon it, a concept generally called *inertia*.
- Newton's Second Law of Motion defines the relationship between acceleration, force, and mass.
- Newton's Third Law of Motion states that any time a force acts from one object to another, there is an equal force acting back on the original object. If you pull on a rope, therefore, the rope is pulling back on you as well.

---

[*] Wikipedia, http://en.wikipedia.org/wiki/space, p. 4.

Let's step a little back in history and look into the roots of the theories that were developed in the 17th century. Many historians claim that the *Tusi-Couple* developed by *Nasir al-Din al-Tusi* who was born in Tus-khurasan, Persia, on February 17, 1201, and died in Baghdad on June 25, 1274, was used by Copernicus after he discovered it in al-Tusi's work.[1] In his treatise,[2] Nasir gave a new model of lunar motion, essentially different from Ptolemy's. In his model Nasir, for the first time in the history of astronomy, employed a theorem invented by himself which, 250 years later, occurred again in Chapter IV of Book III of Copernicus' "De Revolutionibus" (on the revolution of the heavenly spheres). This theorem runs as follows: "If a point moves with uniform circular motion clockwise around the epicycle while the center of the epicycle moves counterclockwise with half this speed along an equal deferent circle, the point will describe a straight-line segment."

The Philosophy of Space and how it was developed through time is a detailed and well documented one, to very briefly point out the major milestones in this centuries-long journey however, one can highlight the followings:

In the *early 11th century,* Islamic philosopher and physicist, *Ibual Haytham* (also known as *Alhacen* or *Alhazen*), discussed space perception and its epistemological implication in his *Book of Optics* (1021). *Abu Ali al-Hasan ibn-al Haytham* was born in Barsa which was then a part of Persia. This Persian scientist and philosopher provided experimental proof of the intromission model of vision which led to changes in the way the visual perception of space was understood, contrary to the previous emission theory of vision, supported by *Euclid* and *Ptolemy*.

In the *17th century*, the philosophy of space and time became the central issue in epistemology and metaphysics. *Gottfried Leibniz*, the German philosopher-mathematician, and *Isaac Newton*, the English physicist-mathematician, discussed two opposing theories for what constitutes space. From *Leibniz's* point of view, space was an idealized abstraction from the relations between individual entities or their possible locations and therefore could not be continuous but must be discrete. *Newton*, on the other hand, took space to be more than relations between material objects and based his position on observation and experimentation and argued that space must exist independently of matter.

In the *18th century*, German philosopher *Immanuel Kant* developed a theory of knowledge in which knowledge about space can be both a *priori* and "synthetic." According to Kant's theory, space and time are not discovered by humans to be objective features of the world but are part of an unavoidable systematic framework for organizing our experiences. Kant rejected the view that space must be either a substance or a relation.

In the *19th century, Carl Friedrich Gauss,* another German mathemati-
cian, considered an empirical investigation of geometrical structure
of space for the first time. In 1905, *Albert Einstein* published a paper on
a "special theory of relativity," in which he proposed that space and
time be combined into a single construct known as "space-time." In
this theory, the speed of light in a vacuum is the same for all observers, which has the result that two events that appear simultaneous to
one particular observer will not be simultaneous to another observer
if the observers are moving with respect to one another. Later
Einstein worked on a "general theory of relativity" that is a theory of
how gravity interacts with space-time. According to this theory "time
goes more slowly at places with lower gravitational potential" and
"rays of light bend in the presence of a gravitational field."

2.2 Outer Space

As mentioned earlier, outer space refers to any location outside the Earth's
atmosphere.

By the National Aeronautics and Space Administration's (NASA) description, the Earth's atmosphere is a blanket of air surrounding the Earth and
reaches over 560 km (348 miles) from the surface of the Earth. The envelope
of gas surrounding the Earth changes from the ground up, and four distinct
layers have been identified using thermal characteristics, chemical composition, movement, and density. Figure 2.1 shows the layers of atmosphere.
There is a layer called the exosphere that starts at the top of the thermosphere
and continues until it merges with interplanetary gases, or space. This layer
continues until about 10,000 km above the surface of the Earth. In this region
of the atmosphere, hydrogen and helium are the prime components and are
only present at extremely low densities. In February 2008, the Conference on
Disarmament agreed that the term "outer space" means the space above the
Earth in excess of 100 km above sea level.

2.2.1 Outer Space Laws

Space law refers to the law that encompasses national and international laws
governing all aspects of outer space activities. Recent history of space law
began with the launch of the world's first artificial satellite, Sputnik, by the
Soviet Union in October 1957.

Space has an infinite number of resources and to set laws regarding these
resources, if not impossible, would be very difficult. With the goal of setting
outer space firmly in place and the nation mobilized around the common
sociopolitical goal of defeating the Russians in the Cold War, an assembly of

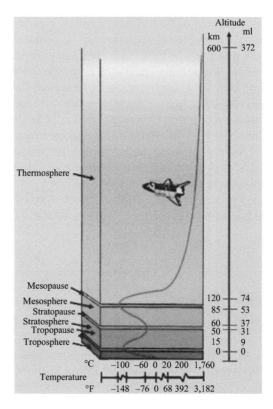

FIGURE 2.1
Layers of atmosphere. (NASA, www.nasa.gov/audience/forstudents/9–12/features/912_liftoff_atm.html.[3])

National Security and States Department officials sought to quash President John F. Kennedy's plans and eliminate this inspirational force.[4] The 1967 Space Treaty prohibits any country from asserting sovereignty over any celestial body, thereby eradicating global international rivalry as a key ingredient in space exploration. In 1958, shortly after the launch of Sputnik, the General Assembly of the United Nations (UN) established an *ad hoc* Committee on the Peaceful Uses of Outer Space (COPUOS) by Resolution 1348 (XIII).[5]

2.2.2 United Nations Committee on the Peaceful Uses of Outer Space

The purpose of this Committee, which had 18 members at the time of its formation, was to consider the following:

- All of the activities and resources of the UN, its specialized agencies, and other international organizations regarding the peaceful uses of outer space.

- International cooperation and programs in related fields that could be restricted so that they all fall under the United Nations Committee on the Peaceful Uses of Outer Space (UNCOPOUS).
- Organizational arrangement for such a plan.
- Legal and regulatory issues that relate to the COPUOS.

On December 12, 1959, the General Assembly established a COPUOS as a permanent body of the UN through Resolution 1472 (XIV).[6] On December 20, 1961, the General Assembly of the UN in its 1058th plenary meeting adopted Resolution 1721 (XVI)[7] through which member countries conducting space launches are requested to provide the information on their launching to the UN. Following this resolution, the "Convention on Registration of Objects Launched into Outer Space," which was adopted by the General Assembly in 1974, entered into force in 1976.[8] The status of an international agreement relating to activities in outer space as of January 1, 2010, is reflected in the Addendum to the "United Nations treaties and principles on outer space and related general assembly resolutions." This report is available at the end of this chapter.[9] On December 13, 1963, the General Assembly of the UN in its 1280th plenary meeting adopted Resolution 1962 (XVIII).[10] In 1966, the Legal Subcommittee of the COPUOS (COPUOS is made up of two subcommittees: the Legal Subcommittee and the Technical Subcommittee) considered the Outer Space Treaty. In the General Assembly, in the same year an agreement was reached which is reflected in Resolution 2222 (XXI).[11] For details of this resolution refer to the end of this chapter. This treaty was based on Resolution 1962 (XVII), adopted in 1963, as was referred to above, with the addition of a few provisions. The treaty entered into force in October 1967. The Outer Space Treaty provides the basic framework on international space law.[12] Glen H. Reynolds and Robert P. Merges[13] explain the problems of law and policy in outer space. In their words, what has been lacking is a comprehensive introduction to the subject. That is the purpose of this book. The book is organized around the needs of space industry and those who deal with it. All private companies become involved in providing launch services and in conducting research and manufacturing in outer space, and as governments together for multinational space stations and other ventures, legal questions are raised that touch on virtually every subject in the law school curriculum, but in new settings (p. xv).

The authors refer to some potential legal issues relating to outer space.[14] For example, they point to often-asked questions: "how far up into the sky over their territory should nations be able to exert legal control?" (pp. 38–40). "What should be done about the increased crowding of geosynchronous orbit?" (pp. 214–215). "Can nations claim sovereignty over celestial objects or particular orbits around the Earth?" (pp. 41, 69–70). Issues such as the above have been addressed by organizations like the International

Telecommunications Union (ITU) and some resolutions have been adopted by member countries.

For example, the World Administration Radio Conference (WARC) convened two meetings in 1985 and 1988, which the author attended, by ITU addressing the use of geostationary satellite orbit (GSO) for radio services. Both conferences were aimed at resolving major policy issues in the international regulation regarding the fair use of GSO by all member countries. The first session, WARC-ORB (1), met in Geneva in August/September 1985, and major changes were agreed for the regulations of the fixed satellite services. The main purpose was to guarantee access for all ITU member nations to the GSO in certain frequency bands and assure the efficient use of other frequency bands. The second session of the conference addressed the orbit-frequency allotment plans for the bands to which access is guaranteed. As far as the GSO orbit allocation was concerned, before the WARC-ORB conference, the GSO could be used by any country that had the technology and was able to launch a satellite into the GSO orbit. A major part of the conference was spent on making sure of an equitable allocation of GSO orbit to all member countries that had plans to launch a satellite into that orbit. After a very long and at times heated discussions between the member countries regarding resolution of this concern, the conference arrived at a conclusion that was based on fair and equitable access and utilization of the GSO-Arc.

The conclusion arrived at was the allocation of parts of the GSO-Arc to each of the member nations. Of course, there were conditions attached to this resolution. The country to which the orbital location or locations were allocated had to comply with ITU regulations in terms of showing sufficient evidence that demonstrated the country's plans for launching a satellite, or satellites, into the allocated orbital slots in the GSO orbit. Such orbital positions allocation was not on an indefinite basis either. If the time allowed to a country to launch its satellite would approach and no evidence was provided showing the intention to launch a satellite into orbit, then ITU would notify the country. If ITU does not receive a definite plan from the country, it would then reallocate the orbital position to another country that was in need and did not have any available orbital slots left.

This plan was fair for both developing and developed nations and would give opportunities to all nations to take advantage of their allocated orbital positions. There were cases that the country could not launch its own satellite into the orbital location allocated to them within the permitted time frame given by ITU and the country had to buy a "used satellite" from another country or organization and move the satellite to its orbital location to make sure its position in the GSO is not taken away. Although this was not a general practice, several countries took advantage of this opportunity.

Surprisingly, there are areas of terrestrial law that could be applied to outer space. International law, communication law, commercial law, intellectual property law, and international trade law are among such laws that Outer Space could benefit from. One needs to note that the international laws are enforced based on treaties that are agreements of the member country to follow the laws adopted by the international organization to which they are members.

The successful implementation and application of the international law governing space activities depend largely on the understanding, accepting, and implementing of such laws by adopting domestic policies that assure such activities. The UN General Assembly convenes its yearly meeting, reaffirming the importance of international cooperation in meeting the space law agreed upon by member nations and urges states that have not yet become parties to the treaties governing the exploration and use of outer space to give serious consideration to ratifying or acceding to them, as well as incorporating them into the national legislations. Other than the yearly meeting of the COPUOS that takes place on a global basis, there are national as well as regional workshops arranged to address this issue. At the time of writing this book, there is a call for participation and contribution to a workshop that is cosponsored by the UN, Thailand, and the European Space Agency (ESA).[15] The objectives of this workshop were as follows:

- To promote understanding, acceptance, and implementation of UN treaties and principles on outer space.
- To promote exchange of information on national space legislation and policies for the benefit of professionals involved in national space activities.
- To consider trends and challenges to international space law, such as commercialization of space activities and an increase of actors involved in space activities.
- To consider development of university-level studies and programs in space law, with a view to promoting national expertise and capability in the field.
- To consider mechanisms for increasing cooperation in the peaceful uses of outer space.

2.2.3 Outer Space Exploration

Exploring outer space has always been a challenge for humans who were curious to find what lies above and beyond our immediate universe. The question as to whether there is life out there or aliens living on another planet, has manifested itself in the form of poems, stories, paintings, or other forms of art and literature. Once astronomy and space technology matured, then space exploration extended to a new phase of getting closer to reality. The use of astronomy and space technology to explore outer space is considered to be a new start in space exploration. Physical space exploration conducted by human space flights and robotic spacecraft became a reality when the development of large and relatively efficient rockets allowed such exploration during the early 20th century. Although space exploration is commonly thought of as advancing scientific research with the goal of ensuring the future survival of humanity and uniting different nations, in reality it was a starting point in human history as a tool for military and strategic advantages of one country relative to others, and hence space exploration faced various criticisms.

In 2006, 14 space agencies came together to begin a series of discussions on global interests in space exploration. They took the unprecedented step of elaborating a vision for peaceful robotic and human space exploration, focusing on destinations with the potential to sustain human habitation within the Solar System, and developed a common set of key space exploration themes. This vision was articulated in "The Global Exploration Strategy: The Framework for Coordination," released in May 2007.

A key finding of this framework document was the need to establish a voluntary, non-binding international coordination mechanism, namely the International Space Exploration Coordination Group (ISECG),[16] 1 through which individual agencies may exchange information regarding interests, objectives, and plans in space exploration with the goal of strengthening both individual exploration programs as well as the collective effort.

The ISECG is important from a policy perspective because it is product-oriented. Its products such as exploration architectures and supporting documents reflect a consensus among the world's space agencies. For example, ISECG spent almost a year developing and reaching consensus on a Reference Architecture for Human Lunar Exploration, which was released in July 2010.[17] This ISECG architecture was developed in response to plans announced by NASA following the announcement by President Bush of a US Vision for Space Exploration for permanent human presence on the Moon in January 2004. The initial NASA architecture focused on a single lunar base but as a result of the growing ISECG consensus, NASA revised its plans to reflect a more distributed lunar surface architecture.

When President Obama cancelled the US program to return to the Moon, ISECG began the development of a Global Exploration Roadmap containing a more accurately reflected global viewpoint for human space exploration. The first version of the Global Exploration Roadmap was issued in September 2011.[18] It is now being used by space agencies to better formulate their plans and to identify areas for cooperation.

We all remember or have heard of the "Space Race" between the old Soviet Union and the United States that started by the launch of the first man-made object to orbit the Earth, the Union of Soviet Socialist Republic's (USSR) Sputnik 1 on October 4, 1957. This challenge was followed by the first Moon landing by the American Apollo 11 on July 1969. If we look at what took place before the space race began, we notice the advances that German scientists made during World War II which laid the foundations for the space race.

The V-2 rocket was the first man-made object that was put into space on October 3, 1942, with the launching of V-4. This led to the first scientific exploration from space which was the cosmic radiation experiment launched by the United States on a V-2 rocket on May 10, 1946.

These space activities in both the United States and USSR prepared the way for the launch of an 83 kg (184 pounds) unmanned satellite called Sputnik that orbited Earth at a height of 250 km (150 miles). Sputnik had two radio transmitters operating at 20 MHz and 40 MHz that transmitted a "beep" signal that could be heard by radios worldwide. Sputnik burned up upon reentry on January 3, 1958. On January 31, 1958, the United States launched Explorer 1 into orbit on a Juno rocket. Explorer 1 discovered the existence of the Van Allen Belts around Earth.

The space race continued with the first human flight by Vostok 1 taking Yuri Gagarin, the 27-year-old Russian cosmonaut, in orbit around the globe in about 108 min on April 12, 1961. This was the start of a new era in space exploration which we can call the "human spaceflight era." The next orbital flight around the Earth was John Glenn's Mercury-Atlas 6 which took place on February 20, 1962. It took 42 years from the Vostok launch for a third country to enter into the human spaceflight race. China launched Yang Liwei into space on October 15, 2003, aboard the Shenzhou 5 (Spaceboat 5) spacecraft.

After successful launch of unmanned satellites and human space flights that orbited the Earth, it was now time for human curiosity to take another step toward outer space exploration. This was the beginning of planetary explorations. The first artificial object ever to reach another celestial body was Luna 2, originally named "Second Cosmic Rocket."

On September 14, 1959, Luna 2 became the first spacecraft to reach the surface of the Moon. It took another decade to reach a new and higher level in space exploration. On July 20, 1969, the first manned landing on another celestial body was performed by Apollo 11 in its lunar landing. This was followed by interplanetary missions. In 1970, Venera 7 landed on Venus collecting and returning data to Earth for 23 min.

Exploring outer space and its potentials to help humans to survive is a thought that has been with us since the inception of space exploration. Stephen Hawking, the renowned British theoretical physicist said, "I don't think the human race will survive the next thousand years, unless we spread into space. There are too many accidents that can befall life on a single planet. But I'm an optimist. We will reach out to the stars."[19] Arthur C. Clarke (December 16, 1917–March 19, 2008), who is considered the father of satellite communication and who described geostationary (here used interchangeably with geosynchronous) Earth orbit (GEO), presented a summary of motivations for human exploration of space in his nonfiction semi-technical monograph *Interplanetary Flight*.[20] Clarke argued that humanity's choice is essentially between expansion off the Earth into space, versus cultural (and eventually biological) stagnation and death.[21]

Professor Hawking, Lacasian Professor of Mathematics at Cambridge, said that the method of space exploration and colonization, apparently the stuff of science fiction, could be one possible escape from the human predicament.

A timeline of solar system exploration ordered by date of spacecraft launch was compiled on Wikipedia.[22]

The data show all spacecraft that have left Earth's orbit for the purpose of solar system exploration, including lunar probes. The timeline begins with the launch of Sputnik 1 on October 4, 1957, which was the Russian's first orbiter. Data are categorized for each decade and the last event shown is PICARD, launched June 15, 2010, as a solar orbiter. The planned timeline lists space exploration activities through the 2030s during which the United States will have manned landing on Mars. Space exploration entails regulatory concerns that the UN has addressed in the form of treaties, as briefly mentioned above. The 1967 Outer Space Treaty (OST) "Treaty on Principles Governing the Activities of States in the Exploration and Use of Outer Space, including the Moon and Other Celestial Bodies"; the treaty was adopted on December 19, 1966, by the UN General Assembly through resolution 2222 (XXI).[23]

Following this treaty, the agreement on the Rescue of Astronauts, the Return of Astronauts and the Return of Objects launched into outer space (1968 ARRA), the International Liability for Damage Caused by Space Objects (1972 LIAB), Registration of Objects Launched into Outer Space (1975 REG), and International Telecommunication Constitution and Convention (1992 ITU), were part of a series of treaties, agreements, and conventions that addressed space exploration and its multidimensional impact on humans worldwide. The most recent discussion on outer space exploration took place at the UN General Assembly COPUOS at its 53rd session in Vienna, June 9–18, 2010.[24] In this meeting, agenda item 6 was the "Implementation of the Recommendations of the Third United Nations Conference on the Exploration and Peaceful Uses of Outer Space (UNISPACE III)." The subcommittee noted the use of telecommunication in the content of tele-health and Earth observation applications in the content of tele-epidemiology, with an emphasis on improving public health and infectious disease management.

At the UNISPACE III conference, held in Vienna (July 9–30, 1999), the report[25] summarized 33 conclusions and proposals on different activities related to the exploration and peaceful use of outer space. Items VI, "Conclusions and proposals of the symposium on recent progress and future plans for exploration of the solar system," Item X "Conclusion and proposals of the symposiums on contribution of space techniques to the exploration of universe," and items XXXIII, "Conclusions and proposals of the workshop on Mars exploration," addressed the issue of outer space exploration in some detail. In the recent progress and future plans for exploration of the solar system, four large space agencies issued reports. Japan reported on the Moon, Russia on Mars, ESA on Mars and Mercury, and NASA on Moon and Mars explorations.

In item X, the Committee on Space Research (COSPAR) organized this activity. The group expressed their satisfaction on the progress made since the second UN conference on the exploration and peaceful uses of outer space, held in 1982.

In Item XXXIII, which focuses on Mars exploration, a Lidar instrument developed by the Russians to measure atmospheric dust and haze was mentioned. In addition, Ariave-5 was reported to deliver micromissions including four NETLANDERs, to the surface of Mars, which would study the interior of the planet and further track the evolution of water on it.

As I was searching through the literature on space exploration, I came across a 1914 patent by *Robert H. Goddard*[26] detailing a rocket apparatus. In this patent, a mechanism for bringing back the rocket to Earth without damage was described. It is fascinating to see that a series of rockets, all within the same structure, are designed such that more rockets could be discharged to continue the mission to any desired extent. (Refer to the end of this chapter for the details of this patent).

Because of such visionary scientists, humans have achieved what we witness today.

2.2.4 Outer Space Challenges Ahead

The main challenge that needs to be globally accepted and plans developed accordingly, is how to use space for the benefit of the Earth and its entire people. Secure World Foundation's President and Co-founder Cynda Collins Aresenault[27] asks: "What new leaders will step forward to inspire us in terms of our use of space for problem solving?" "What new discoveries will be made?" "What new inventions will open doors?" If one were to point to some of the challenges ahead in outer space, the following would be most probably among them:

- International agreement on a Code of Conduct for outer space activities
- Sustainability of outer space
- Space traffic management
- Space security

In the following sections, each of the above concerns will be addressed, but before we embark on addressing these crucial issues, let's look at some of the visions and challenges that lie ahead. Future visions for information technology (IT) in space could be summarized as follows:

- Software-based capabilities enable space platforms with new onboard capabilities and long-term survivability.
- The role of software, IT, and computing in space mission success is known and valued.

With no doubt, a future explosion of raw data collected by space platforms will require more capable, more reliable, and more complex software engineering to analyze such data. Modern information, software engineering, and computing technologies have numerous critical contributions to make toward the success of future space missions. The creation of sophisticated prediction models to define scientific investigations; the systematic exploration and detailed understanding of mission and space platform design; new approaches to the validation of complex software; capabilities for mission planning and execution in remote planetary environments; architectures for coordination of multiple space platforms; scaled bandwidth and networking to bring much more information from deep space back to Earth; and new ways to assist scientists searching for the nuggets in the data, even at collection time, are among so many challenges to be met.[28]

Roger Launius, chair of the Division of Space History of the National Air and Space Museum, suggested five major challenges that need to be made in future space exploration[29]:

- Political will
- Cheap, reliable access to space
- Smart robots
- Protection of the planet and the species
- Exploration beyond low-Earth orbit

Bob Preston and *John Baker* in "Space challenges"[30] begin the discussion on challenges ahead in space by explaining how space activities are divided in the United States. Civil, national security, and commercial are mentioned as three distinct space activities. Civil sector refers to research and development activities, national security covers intelligence activities, and commercial sector refers to communications, remote sensing, launch, and so forth.

2.2.4.1 International Agreement on a Code of Conduct for Outer Space Activities

A code of conduct for outer space activities should be able to "shape a set of principles for respecting the rights of all spacefaring nations and users of Satellites to operate in space," says Williamson of the Secure World Foundation.[28] Michael Krepon, who is a leader in advocating a code of conduct for space, notes that collaboration in drafting a code is underway with nongovernmental organizations (NGOs) within countries to internationalize the affect.[31]

2.2.4.2 Sustainability of Outer Space

Sustainability of space activities has been a matter of concern for several decades especially for space-faring nations, regional space organizations, and commercial satellite operators. Countries not in a position to independently utilize space because of their technology and obstacles have also been concerned about such activities by those who can utilize the space, from the security point of view. Space activities are not limited to commercial applications but rather encompass spy satellites, remote-sensing satellites, and so on. These space-borne technological tools are capable of collecting information from the surface as well as underground activities and resources globally which to some interpretation violates the sovereignty of nations. Legal background for the long-term sustainability of space activities is found in several instruments and provisions of international law as well as national legislation dealing with concerns about the sustainable future of space activities. The 1967 UN Outer Space Treaty, as complemented by the 1972 UN Liability Convention, states that the launch or procurer of a launching or from whose territory or facility an object is launched are internationally liable for any damages caused by this space object on the Earth's surface and to aircraft in flight as well as for damage caused by their fault to another space object in flight.

The Inter-Agency Space Debris Coordination Committee (IADC) focuses on the proliferation of space debris by the IAA (International Academy of Astronauts) or by IAASS (International Association for Advancement of Space Safety). A major threat to the long-term sustainability of space activities is the growing rate of space debris population resulting from many low-Earth-orbit (LEO) satellite constellations. The space debris includes all man-made objects such as nonoperational spacecraft, derelict launch vehicle upper stages, and fragments from satellites that are in orbit around the Earth. As of July 2008,[32] the number of orbital debris greater than 10 cm in size were 18,000 and those of greater than 1 mm in size estimated to be more than tens of millions. The impact of even the small objects in space is huge due to the high relative velocities they have (1–14 km/s depending on the orbit). The collision hazard to operating satellites especially in the lower orbits is a growing concern in the satellite industries. Because of such concern, the UN established Space Mitigation Guidelines through its Technical Subcommittee of UNCOPUOS:

- Limit debris released during normal operation.
- Minimize potentials for break-ups during operational phases.
- Limit the probability of accidental collision in orbit. Avoid intentional destruction and other harmful activities.
- Minimize the potential for post-mission break-ups resulting from stored energy.
- Limit the long-term presence of spacecraft and launch vehicle orbital stages in LEO after the end of their missions.

Among the above mitigation measures, the most effective ones are the implementation of end-of-life disposal maneuvers and end-of-life passivation of spacecraft and orbital stages. Some of the recommendations made by the UN toward the improvement of the safety and security of space activities are as follows:

- Compliance with and promotion of treaties, conventions, and other commitments relating to outer space activities.
- Following of common and universal protocols for space communications and data standards among space operation.
- Provision of information about the location as well as timing of planned maneuvers of satellites, particularly in transfer orbit missions to other GEO operations.
- Coordination of spacecraft in adjacent orbital location by different operator entities.
- Exploration of mechanisms to coordinate the tracking of hazardous space objects and to coordinate alert processes for those who are operating spacecraft.
- Improvement of the international guidelines to ease their implementation by the operator.

Figure 2.2 illustrates the increasing number of catalogued space objects from 1957 to the end of 2008, most of them being space debris.

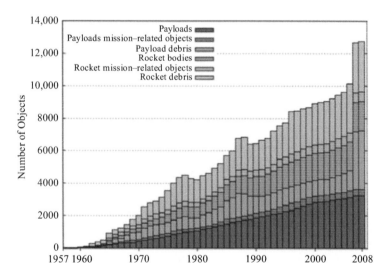

FIGURE 2.2
Evolution of the space objects population. (Presentation by Heiner Klinkrad of the European Space Agency to the Scientific and Technical Subcommittee of the UNOPUOS, February 2008.[33])

2.2.4.3 Space Traffic Management

Managing traffic in space is a challenging task, especially when there are numerous satellite constellations in different orbits. Such challenges include collision avoidance, human zone, sun synchronous zoning, and geosynchronous maneuvers. In order to assure safe and secure traffic in space, a set of space traffic management (STM) rules should be in place and followed by those who are participating in space activities. These rules that are not focused on debris mitigations allow more efficient use of crowded orbits and give owner-operators the tools to protect their spacecraft. STM provides standard data set, warnings, and recommendations of avoidance maneuvers to help owner-operators who might not have the tracking or analytical ability in-house. It also gives the owner-operators the flexibility of maneuvering based on internal cost–benefit analysis. These rules provide the spacecraft owner-operators with the information and tools to help make educated choices and to improve satellite safety. The human-rated zoning creates a protected zone for human traffic with minimal impact to current and future non-human-rated operators. STM rules also increase the efficiency of existing GEO slotting and operations and reduce energy costs. Following these rules will allow for more efficient planning for station-keeping maneuvers. Obstacles that might remain in the way of following and implementing such rules are the reluctance to share data due to privacy and competitive advantage concerns and the legitimacy of STM organizational body to implement and enforce rules. Brian Weeden and Ben-Baseley Walker[34] outline the key steps for STM systems in three phases: Phase 1: Develop Rules, Phase 2: Build Consensus, and Phase 3: Implement the systems. The potential STM organizations are shown in Figure 2.3.

Managing Body	Phase I: Rule Development	Phase II: Consensus	Phase III: Implementation of the System (1)	Phase III: Arbitration Procedures (2)
UNCOPUOS		✓		✓
ITU		✓		
IADC	✓			
ICAO			✓	✓
New Agency	✓		✓	✓

FIGURE 2.3
Potential STM organization.

2.2.4.4 Space Security

Some of the most serious threats to space security are environmental. The increasing number of objects in Earth's orbit combined with the fact that they are moving at an extremely high speed can endanger space assets. This concern necessitates continuous monitoring of the rapidly changing space environment in order to preserve the security in outer space. According to reports,[35] there are currently over 19,000 objects 10 cm in diameter or larger that are being tracked by the U.S. Space Surveillance Network, over 90% of which are space debris.

According to the same report the enhancement of Space Situational Awareness capabilities can contribute to more efficiently tracking debris and supporting collision avoidance. Effectively monitoring and tracking objects in outer space are only possible when all the private agencies and governmental organizations actively participate in sharing their data. One obstacle that remains as a concern that is somewhat difficult to tackle is the confidentiality of the information related to space-based military applications. Over 130 space experts from 17 countries in the civil, commercial, and military space sectors have established the Space Security Index to assess space security. This index provides background information and in-depth analysis on the key space security trends and developments of nine indicators of space security.

The Space Security Index is defined as the secure and sustainable access to, and use of space and freedom from space-based threats. The nine indicators of the Space Security Index are as follows:

- The space environment
- Space situational awareness
- Space security laws, policies, and doctrines
- Civil space programs and global utilities
- Commercial space
- Space support for terrestrial military operations
- Space systems protection
- Space systems negation
- Space-based strike weapons

Will Marshall[36] in his presentation titled "Space Security: Ways Forward," lists nine—vulnerabilities (adopted from de Blois et al., *International Security*, Fall 2004) as follows:

T1—Electronic warfare, such as jamming or spoofing of satellite communication

T2—Physical attacks on satellite ground stations, an example of this is the attack of Iraqi's military by air-to-surface missiles that destroyed the only Earth station in Hamadan, Iran, during the 8-year war imposed on Iran by Iraq

T3—Dazzling or blinding of satellite sensors

T4—Radio-frequency (RF) weapons (e.g., high-powered microwaves, HPM)

T5—"Heat-to-kill" ground-based laser anti-satellite (ASAT) weapons

T6—Pellet cloud attacks on low-orbit satellite

T7—Attacks in space by microsatellites

T8—Hit-to-kill ASAT weapons

T9—High-Altitude Nuclear Detonations (HANDs)

Some experts at NASA believe that collision between space assets and larger pieces of debris will remain rare only for the next decade, although there are still ongoing discussions on this assessment. Figure 2.4 shows the top 10 breakups as of May 2010. Figure 2.5 shows the total on-orbit debris by launching state, and Figure 2.6 shows the monthly number of objects in Earth's orbit, by type.[37]

Figure 2.7 shows the classification of geosynchronous objects. In addition to treaties, five UN resolutions known as UN principles have been adopted by the UN's General Assembly regarding space activity categories. Figure 2.8 shows these key UN space principles which are the code of conduct reflecting the conviction of the international community on the issues.

Common Name	Year of Breakup	Altitude of Breakup	Cataloged Debris*	Debris in Orbit*	Cause of Breakup
Fengyun-1C	2007	850 km	2841	2756	Intentional Collision
Cosmos 2251	2009	790 km	1267	1215	Accidental Collision
STEP 2 Rocket Body	1996	625 km	713	63	Accidental Explosion
Iridium 33	2009	790 km	521	498	Accidental Collision
Cosmos 2421	2008	410 km	509	18	Unknown
SPOT 1 Rocket Body	1986	805 km	492	33	Accidental Explosion
OV 2-1 / LCS 2 Rocket Body	1965	740 km	473	36	Accidental Explosion
Nimbus 4 Rocket Body	1970	1075 km	374	248	Accidental Explosion
TES Rocket Body	2001	670 km	370	116	Accidental Explosion
CBERS 1 Rocket Body	2000	740 km	343	189	Accidental Explosion
*As of May 2010			Total: 7903	Total: 5172	

FIGURE 2.4
Top 10 satellite breakups. (Marshall, Will, 2008, Space Security Ways Forward, SETI Institute, California, October 12.[36])

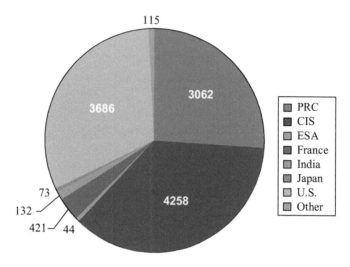

FIGURE 2.5
The total on-orbit debris by launching state, 2009.

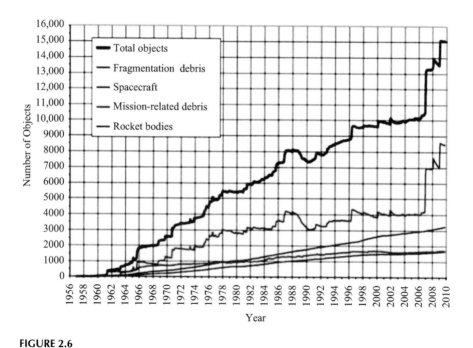

FIGURE 2.6
The monthly number of objects in Earth's orbit, by type. (Tabulation of "Orbital Box Score Data," *Orbital Debris Quarterly*, January 2010, www.orbitaldebris.JSC.nasa.gov/newsletter/pdfs/ODQNv14i1.pdf.[37])

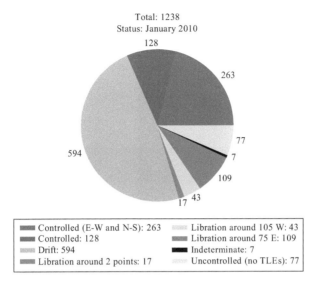

FIGURE 2.7
Classification of geosynchronous objects. (Tabulation of "Orbital Box Score Data," *Orbital Debris Quarterly,* January 2010, www.orbitaldebris.JSC.nasa.gov/newsletter/pdfs/ODQNv14i1.pdf.[37])

Declaration of Legal Principles Governing the Activities of States in the Exploration and Users of Outer Space (1936)
Space exploration should be carried out for the benefit of all countries.
Outer space and celestial bodies are free for exploration and use by all states and are not subject to national appropriation by claim sovereignty.
States are liable for damage caused by spacecraft and bear international responsibility for national and nongovernmental activities in outer space.
Principles on Direct Broadcasting by Satellite (1982)
All states have the right to carry out direct television broadcasting and to access its technology, but states must take responsibility for the signals broadcasting by them or actors under their jurisdiction.
Principles on Remote Sensing (1986)
Remote sensing should be carried out for the benefit of all states, and remote sensing data should not be used against legitimate rights and interests of the sensed state.
Principles of Nuclear Power Sources (1992)
Nuclear power may be necessary for certain space missions, but safety and liability guidelines apply to its use.
Declaration on Outer Space Benefits (1996)
International cooperation in space should be carried out for the benefit and in the interest of all states, with particular attention to the needs of developing states.
UN Space Debris Mitigation Guidelines (2007)
Voluntary guidelines for the mission planning, design, manufacture, and operation phases of spacecraft and launch vehicle orbital stages to minimize the amount of debris created.

FIGURE 2.8
Key UN space principles. (SpaceSecurity.org, 2010, Library and Archives Canada Cataloguing in Publications Data, Space Security.[35])

Although the United States and Russia started the exploration of space in the 1950s, in the following decades a handful of other countries developed their own independent orbital launch capabilities. Figure 2.9 shows these countries and their launching capabilities through 2009.

Figure 2.10 shows worldwide orbital launch events in 2009, and Figure 2.11 shows the international space security-related institutions, which are UN-related institutions.

State/Actor	Year of First Orbital Launch	Launch vehicle	Satellite
USSR/Russia	1957	R-7 rocket	Sputnik 1
USA	1958	Jupiter-C	Explorer 1
France*	1965	Diamant	Astérix
Japan	1970	Lambda	Osumi
China	1970	Long March	Dong Fang Hong I
UK*	1971	Black Arrow	Prospero X-3
India	1980	SLV	Rohini
Israel	1988	Shavit	Ofeq 1
Iran	2009	Safir-2	Omid

FIGURE 2.9
Countries with independent orbital launch capacity. * France and the United Kingdom (UK) no longer conduct independent launches, but France's CNES manufactures the Ariane launcher used by Arianespace/ESA.

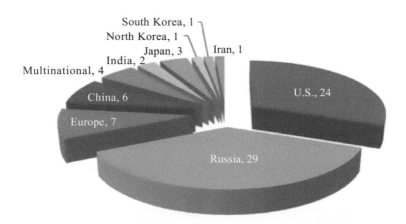

FIGURE 2.10
Worldwide orbital launch events in 2009. The launch attempts of North Korea and South Korea were not successful and their respective payloads were not placed in orbit.

Source: Space Security.org, 2010, Library and Archives Canada Cataloguing in Publications Data, Space Security.[35]

FIGURE 2.11
International space security—UN-related institutions. (SpaceSecurity.org, 2010, Library and Archives Canada Cataloguing in Publications Data, Space Security.[35])

2.3 Space within Reach

In this section, the space within reach is referred to as the space activities that are carried out at such an altitude that one could access for repair, delivery, and research on board with human presence. In the following sections, the space shuttle program, Buran—the Soviet Shuttle, the Mir Space Station, the International Space Station, and space activities in some of the other countries are explained.

2.3.1 Launch Systems

Launch systems are generally categorized into two distinct groups. The first category which is the traditional method of launching and is still the dominant method for launching satellites into low, medium, or high orbits and the second category are the reusable launching vehicles that are less common but are used for launching satellites mostly into the low orbits. In what follows, we will discuss both methods and their applications.

2.3.2 Conventional Launch Systems

Launch vehicles have the primary function of putting a spacecraft into an orbit or a suborbital trajectory. Most of the rocket stages fall away in sequence in the process of launching a satellite until the spacecraft reaches its planned orbit. In a conventional launch process, the first stage of the rocket propels the rocket from the launch pad, and then the second-stage rocket boosts the payload to orbit. Launch vehicles generally are used only once, although there are reusable launch vehicles that will be discussed in the next section. Figure 2.12 is a cross-section of a typical launch vehicle, the United Launch

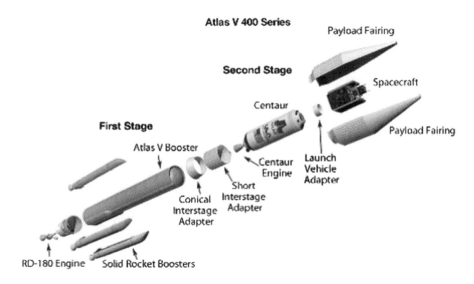

FIGURE 2.12

A typical arrangement for a launch site and range. Cross Section of a Typical Launch Vehicle.

Source: **Image courtesy of United Launch Alliance, http://www.ulalaunch.com/products_ atlasv.aspx.**

Alliance Atlas V.[38] Its major components—from top to bottom—include the following (Figure 2.13):

- The *nose cone or fairing*, a structure made with a vented aluminum-honeycomb core and graphite epoxy covering, carries the payload. Manufacturers offer clients a choice of three payload fairings, depending on the size of the payload. The fairing protects the payload from atmospheric pressure changes and aerodynamic heating during launch.

- The *second-stage rocket* consists of fuel and oxygen tanks, control systems, and a rocket engine that carries the payload to orbit. The Atlas V is propelled by a single RL 10 Centaur engine and stainless steel fuel tanks, providing 22,300 pounds-force (lbf) of thrust, fueled by liquid hydrogen and liquid oxygen. The walls of the stainless steel tanks are insulated and so thin that they cannot support their own weight before they are pressurized; a design developed to maximize engine performance.

- *Adapters* connect the first and second stages of the rocket and provide the structure for housing vehicle electronics.

- The *first stage* consists of additional fuel and oxygen tanks, control systems, and rocket engines, sometimes supplemented with strap-on boosters. The Atlas V main booster is made of a special aluminum

Launch System	Upper Stage	LEO (kg)	GTO (kg)	GEO (kg)	Polar (kg)
Atlas IIAS	Centaur 2A	8640	3606	1050	7300
Delta II 7920/25	PAM-D	5089	1840	910	3890
Pegasus XL		460			345
Shuttle	- IUS	24.400	5900	2360	
Taurus	Star 37	1400	450		1060
Titan IV	- Centaur		8620	4540	14.110

FIGURE 2.13
Examples of launch systems for different orbits.

and, unlike the stage-two tanks, is structurally stable. The launch vehicle is fueled by rocket propellant (or highly purified kerosene) and liquid oxygen that provide 860,300 lbf of thrust. The RD 180 engine was developed in Russia and is produced by a U.S.–Russian joint venture.

2.3.2.1 General Description

A typical launch vehicle system consists of several basic subsystems, including propulsion; power; guidance, navigation, and control (GNC); payload adapters; and fairings. In the followings, the focus is put on the propulsion subsystem, specifically rocket engines used on U.S. launch vehicles.

2.3.2.2 Rocket Engines

Rocket engines are generally grouped according to the type of propellant being used: solid or liquid. There are also examples of hybrid engines that feature both solid and liquid propellants.

Rocket engines that burn solid propellant are simpler in construction, relatively inexpensive, and can be stored for long periods of time, making them ideal for missiles in particular. Once ignited, engines burning solid propellant cannot be throttled at will or shut off. These characteristics make it a potentially controversial option for launch systems designed to carry people.

The following solid motors are used in currently available U.S. launch vehicles:

- **Five-Segment Solid Rocket Boosters (SRB):** The five-segment SRB is derived from the four-segment SRBs used for Space Transportation System (STS) from 1981 to 2011.
- **STAR motors:** The STAR line of solid motors, first produced by Thiokol and now manufactured by Orbital ATK, is used for upper stage elements in launch vehicles.
- **GEM Strap-on Booster System:** The Graphite Epoxy Motor (GEM) provided by Orbital ATK was introduced in 1991 to supplement the first-stage thrust of the Delta II launch vehicle.
- **AJ-60A Solid Rocket Motor:** The AJ-60A solid motors, manufactured by Aerojet Rocketdyne, have been used to supplement first stage-thrust for the Atlas V since 2002.

The following liquid rocket engines are used in currently available U.S. launch vehicles. Also included are engines designated for use on vehicles under development.

- **BE-4:** The BE-4 is an engine under development by Blue Origin. It will burn a mixture of liquid oxygen (LOX) and liquefied natural gas (LNG), mostly composed of methane) and produce 2,447 kN (550,000 lbf) of thrust. This is the baseline engine for the company's orbital launch vehicles and the first stage of ULA's Vulcan.
- **AR-1:** The AR-1 is an engine currently under development by Aerojet Rocketdyne. The engine, which will burn a LOX-kerosene mixture, is designed to produce about 2,224 kN (500,000 lbf) of thrust.
- **FRE-1 and FRE-2:** Fire Fly Space has developed the FRE line of engines to power the first and second stages of its Alpha launch vehicle. The FRE-2 is an aerospike engine that, if successful, may prove to be the first aerospike engine employed in an operational launch system.
- **Merlin 1D:** The Merlin 1D is the engine used to power both the first and second stages of SpaceX's Falcon 9 and Falcon Heavy launch vehicles. This engine produces about 756 kN (185,500 lbf) of thrust and burns a LOX-kerosene mixture. Nine of these engines power the Falcon 9 first stage (for a total thrust of about 6,806 kN or 1,530,000 lbf) and one is used to power the second stage. The Merlin 1D is a fourth-generation SpaceX engine that traces its lineage to the Merlin 1A that powered the Falcon 1 vehicle.
- **Newton:** The Newton series of engines being developed by Virgin Galactic will power the company's air-launched LauncherOne

vehicle. These engines use LOX and kerosene as propellants. The NewtonThree, which produces 327 kN (73,500 lbf) of thrust, will power the LauncherOne first stage. A NewtonFour engine, producing 22 kN (5,000 lbf) of thrust, will power the second stage to orbit.

- **RD-180:** The RD-180 is a Russian-built engine that powers the Common Core Booster (CCB) of the Atlas V vehicle using a LOX-kerosene propellant mixture. It produces a thrust of about 3,830 kN (860,000 lbf). The engine is built by RD AMROSS (a joint effort between Aerojet Rocketdyne—previously Pratt & Whitney Rocketdyne—and NPO Energomash).

- **RD-181:** The RD-181 is an engine being developed by NPO Energomash for the Antares vehicle built and offered by Orbital ATK. The original Antares, which was used on four missions, used two AJ26 engines on its first stage. The AJ26 was essentially a significantly modified NK-33 engine.

- **RL10:** The first variant of the RL10 engine was designed in 1959 by Pratt & Whitney (now part of Aerojet Rocketdyne). It was first used in 1962 as the engines for the Centaur upper stage of Atlas missiles converted as launch vehicles. The engine burns LOX-liquid hydrogen and produces a thrust of about 110 kN (25,000 lbf). The current model of this engine, the RL10A-4-2, continues to power the Centaur upper stage for the Atlas V. The RL10B-2 is used for the Cryogenic Upper Stage of the Delta IV vehicle. Further development of the RL10 is underway to support ULA's Advanced Cryogenic Evolved Stage (ACES) for the company's Vulcan launch vehicle.

- **RS-25E:** The RS-25E, built by Aerojet Rocketdyne, is an expendable version of the RS-25, also called the Space Shuttle Main Engine (SSME). Four RS-25E engines will be used for each core stage of NASA's upcoming SLS. Sixteen SSMEs from the retired STS Program have been refurbished and stored for use on four SLS missions, which begin in late 2018. The RS-25E will be used on subsequent SLS vehicles. Each RS-25E will burn a LOX-liquid hydrogen propellant mixture and produce about 2,277 kN (512,000 lbf) of thrust.

- **RS-27A:** The RS-27A is the engine used to power the core stage of the Delta II. Also developed by Aerojet Rocketdyne, the RS-27A burns LOX and kerosene, producing a thrust of about 890 kN (200,100 lbf).

- **RS-68:** Aerojet Rocketdyne also produces the RS-68, a more powerful engine than the RS-27 that burns a LOX-liquid hydrogen propellant mix. From 2002 to 2012, each Common Booster Core (CBC) of the Delta IV was powered by a single RS-68 engine, which produces about 2,950 kN (660,000 lbf) of thrust. An upgraded version of the engine, called the RS-68A, was introduced in 2012 as a replacement to the RS-68. It can produce 3,137 kN (705,000 lbf) of thrust.

- **Rutherford:** Rocket Lab has designed the Rutherford engine for use in the first stage of the company's Electron vehicle. The engine burns a mixture of LOX and kerosene, producing a thrust of about 22 kN (5,000 lbf). Rocket Lab is employing additive manufacturing (3D printing) in the construction of all primary components of the Rutherford, making it a unique example in the industry. 3D printing reduces costs by simplifying the manufacturing process.
- **XR Series:** XCOR Aerospace has been developing engines since 2000, when the company fully integrated the XR-3A2 and XR-4A3 into an EZ-Rocket test aircraft. Currently, XCOR is developing the XR-5K18 engine for the company's Lynx suborbital vehicle. The XR-5K18 burns a LOX and kerosene propellant mixture, producing a thrust of about 13 kN (2,900 lbf).

2.3.2.2.1 Launch Vehicle Integration and Processing

Since there are many different types of launch vehicles, there are many different ways to integrate and launch them. In general, however, vehicle assemblies and subsystems are manufactured in several locations, and then transported via rail, air, or sea to the launch site where the parts come together as a complete launch vehicle. Figure 2.14 illustrates the basic idea using a generic vehicle as an example.

Once the launch vehicle is fully integrated, it is then joined with its payload. This process is called payload integration. The payload will arrive at the launch site from the manufacturing or checkout site to a specialized facility designed to handle the unique needs of the payload. For example,

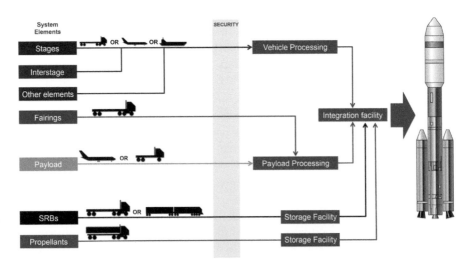

FIGURE 2.14
A typical launch vehicle integration and processing scheme.

payloads may require fueling, last-minute integration with components, or final testing and checkout. It is then attached to a payload adapter. The payload adapter is the physical connection between the payload and the launch vehicle, and can be integrated with the launch vehicle either horizontally or vertically depending on the vehicle. Once integrated, the payload fairing is installed. The vehicle and payload then make their way to the launch pad, where the combination continues to be monitored during a technical checklist called a countdown. Fueling of a vehicle using liquid propellants takes place at the pad, usually immediately prior to launch.

While the launch vehicle and payload are handled at the launch site, other operations take place to support launch activities. These are handled by a launch range, whose main purpose is to insure that the launch is conducted efficiently and safely. The range manages the airspace around the site, any ground or sea traffic in the vicinity or downrange, and supports launch emergencies should they occur.

2.3.2.2.2 Operational Orbital Launch Vehicles

By the end of 2015, there were 112 different orbital launch vehicles operating around the world. This figure includes variants of a family of vehicles; for example, there are 18 Atlas V variants defined by the number of solid rocket boosters used, type of fairing by diameter, and type of Centaur upper stage (single or dual engine). Not all of these vehicles are available for commercial use, whereby a payload customer can "shop around" for a ride into orbit.

There are six expendable launch vehicle types available for commercial use by launch providers in the United States (see Table 2.1). The Delta II, which flew once in 2015, is no longer available. U.S. launch service providers include Maryland-based Lockheed Martin, Virginia-based Orbital ATK, California-based SpaceX, and Colorado-based ULA. ULA has historically only served U.S. government customers, but has indicated it plans to open its Atlas V, Delta IV, and future Vulcan vehicles for international competition. Another U.S. vehicle, the Super Strypi, developed and built by the University of Hawaii (UH), Sandia National Laboratory.

TABLE 2.1

International Space Station (ISS) Characteristics

Crew	Six When Complete
Length	356 ft (108.55 m)
Width	239 ft (72.8 m)
Total mass	927,316 lb (420,623 kg)
Habitable volume	14,400 ft^3 (408 m^3)
Solar array area	32,528 ft^2 (3023 m^2)
Power	80 kW
First assembly launch	November 20, 1998

There are 13 expendable launch vehicle types available for commercial use outside the United States: Ariane 5, Dnepr, Epsilon, GSLV, H-IIA/B, Long March 2D, Long March 3A, Proton M, PSLV, Rockot, Soyuz 2, Vega, and Zenit 3SL/SLB.

2.3.2.2.3 Global Payload Industry

Countries and jurisdictions worldwide that possess functional and operating indigenous payload manufacturing sectors are China, the European Union, India, Japan, Russia, and the United States. Countries that have developed and built their own spacecraft include Argentina, Iran, Israel, North Korea, South Korea, and Ukraine. A total of over 30 countries have developed and built at least one orbital payload, usually a satellite.[39] The payload building capability of more than half of these countries is limited to CubeSats, built from pre-fabricated kits by universities and government and non-profit organizations.

Figure 2.15 presents civil, military, and commercial orbital payloads, by country of manufacturer, in 2015. In 2015, 42 CubeSats, most of them commercial, were launched as cargo for subsequent deployment from the ISS. Sixteen CubeSats for commercial remote-sensing operator Planet Labs were launched aboard Spx-6 mission in April. Eighteen CubeSats were launched in August aboard the HTV cargo spacecraft, 14 of them for Planet Labs and 4 more CubeSats for research and development purposes. Eight Planet Labs' CubeSats were lost in the Spx-7 launch incident in June. These satellites are counted separately because they are not deployed in their respective orbits at the time of the separation from the launch vehicle and are stored inside an on-orbit cargo vehicle.

Country of Manufacturer	Civil	Military	Non-Profit	Commercial	Total
Argentina	0	0	0	1	1
Canada	0	0	0	1	1
China	7	8	24	6	45
Europe	11	1	1	10	23
India	4	0	0	0	4
Indonesia	1	0	0	0	1
Iran	1	0	0	0	1
Japan	1	2	0	1	4
Russia	12	13	0	0	25
Singapore	1	0	4	1	6
South Korea	1	0	0	0	1
USA	18	14	19	33	84
TOTALS	**57**	**38**	**48**	**53**	**196**

FIGURE 2.15
Number of civil, military, non-profit, and commercial, payloads launched in 2015 by country of manufacturer.

2.3.2.2.4 *Commercial On-Orbit Vehicle and Platforms*

NASA started the commercial crew and cargo program to help commercial companies develop new capabilities for transporting crew and cargo to the International Space Station (ISS).[40] These services are intended to replace some of the ISS resupply services performed by the Space Shuttle. The first of these vehicles, SpaceX's Dragon, became operational in 2012, restoring NASA's ability to deliver and retrieve cargo in LEO. Crewed vehicles made many advances in 2015, but are not expected to become operational before 2017.

Boeing continues to develop the CST-100 Starliner and SpaceX is developing the Crewed Dragon for the NASA Commercial Crew Transportation Capability (CCtCap) program. Sierra Nevada Corp., which is developing the *Dream Chaser*, vows to continue working on its winged vehicle.

Figure 2.16 presents civil, military, and commercial orbital launches by country in 2015, and Figure 2.17 shows *total worldwide launch activity during 2015.*

Country/Region	Civil	Military	Commercial	Total
Russia	14	7	5	26
USA	4	8	8	20
China	12	7	0	19
Europe	5	0	6	11
India	3	0	2	5
Japan	1	2	1	4
Iran	1	0	0	1
TOTALS	40	24	22	86

FIGURE 2.16
Total orbital launches in 2015 by country and type.

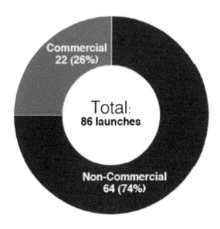

FIGURE 2.17
2015 total worldwide launch activity.

2.3.2.2.5 Vehicle Mass Class

Small launch vehicles are defined as those with a payload capacity of less than 2,268 kg (5,000 pounds) at 185 km (100 nautical miles) altitude and a 28.5-degree inclination. Medium to heavy launch vehicles are capable of carrying more than 2,269 kg at 185 km altitude and a 28.5° inclination.

2.3.2.2.6 Payload Mass Class

Figure 2.18 provides the payload mass classes used by the FAA AST.

Intensive launch: From 2011 till 2016, Chinese LM Launch Vehicle has completed 91 launch missions, 18 launches per year on average and with success rate of 97.8%, both on leading position of the world as shown in the Figure 2.19 below:

Class Name	Kilograms (kg)	Pounds (lb)
Femto	0.01 - 0.1	0.02 - 0.2
Pico	0.09 - 1	0.19 - 2
Nano	1.1 - 10	3 - 22
Micro	11 - 200	23 - 441
Mini	201 - 600	442 - 1,323
Small	601 - 1,200	1,324 - 2,646
Medium	1,201 - 2,500	2,647 - 5,512
Intermediate	2,501 - 4,200	5,513 - 9,259
Large	4,201 - 5,400	9,260 - 11,905
Heavy	5,401 - 7,000	11,906 - 15,432
Extra Heavy	>7,001	>15,433

FIGURE 2.18
Payload mass classes.

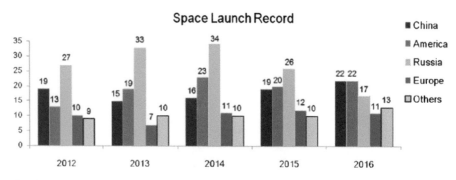

FIGURE 2.19
Space launch record.

2.3.3 Reusable Launch Systems

A **reusable launch system** (**RLS**, or **reusable launch vehicle, RLV**) is a launch system which is capable of launching a payload into space more than once.[41] This contrasts with expendable launch systems, where each launch vehicle is launched once and then discarded.

No completely reusable orbital launch system has ever been created. Two partially reusable launch systems were developed, the Space Shuttle and Falcon 9. The Space Shuttle was partially reusable: the orbiter (which included the Space Shuttle main engines and the Orbital Maneuvering System engines), and the two solid rocket boosters were reused after several months of refitting work for each launch. The external tank was discarded after each flight.[42]

The Falcon 9 rocket has a reusable first stage; several of these stages have been safely returned to land after launch. On March 30, 2017, a reused Falcon 9 successfully landed on an **Autonomous Spaceport Drone Ship (ASDS)**, after its second launch, marking the first successful relaunch and landing of a used orbital-class booster.

Several systems reusing parts of the rocket, such as the first stage of New Glenn and the engine sections of Ariane 6 (Adeline) and Vulcan, are currently under development. One fully reusable system, the Interplanetary Transport System, is also under development.

Orbital RLVs are thought to provide the possibility of low cost and highly reliable access to space. Reusability implies weight penalties such as non-ablative reentry shielding, additional fuel and rocket components necessary for landing, and possibly a stronger structure to survive multiple uses. Given the lack of experience with these vehicles, the actual costs and reliability are yet to be seen.

2.3.3.1 Space Shuttle Program

The space shuttle was developed by the NASA during the 1970s. The space shuttle system consists of four primary elements: an orbiter spacecraft, two SRBs, an external tank to house fuel and oxidizer, and three space shuttle main engines. The shuttle transports cargo into near-Earth orbit 100–217 nautical miles (115–250 statute miles) above the Earth. The space shuttle is launched in an upright position, with thrust provided by the three space shuttle engines and the two SRBs. After 2 min, the two boosters are spent and are separated from the external tank. They fall into the ocean at predetermined points and are recovered for reuse. The space shuttle main engines continue firing for about 8 min. They shut down just before the craft is inserted into orbit. The external tank is then separated from the orbiter. It follows a ballistic trajectory into a remote area of the ocean but is not recovered. Space shuttle specifications are shown in Figure 2.20A and B. The orbiter's velocity in orbit is approximately 25,405 ft/s (17,322 statute mile/h).

(A)

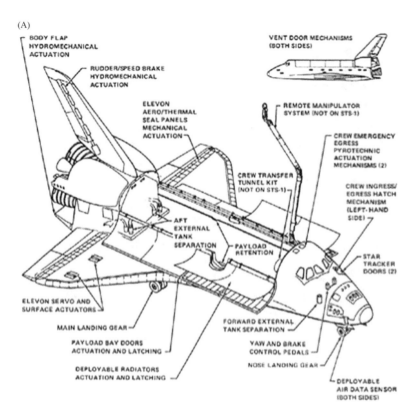

(B)

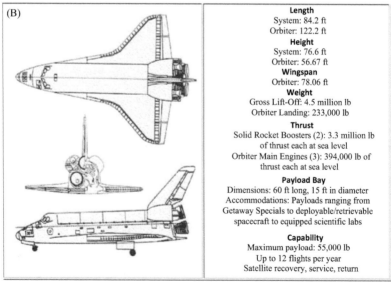

	Length
	System: 84.2 ft
	Orbiter: 122.2 ft
	Height
	System: 76.6 ft
	Orbiter: 56.67 ft
	Wingspan
	Orbiter: 78.06 ft
	Weight
	Gross Lift-Off: 4.5 million lb
	Orbiter Landing: 233,000 lb
	Thrust
	Solid Rocket Boosters (2): 3.3 million lb of thrust each at sea level
	Orbiter Main Engines (3): 394,000 lb of thrust each at sea level
	Payload Bay
	Dimensions: 60 ft long, 15 ft in diameter
	Accommodations: Payloads ranging from Getaway Specials to deployable/retrievable spacecraft to equipped scientific labs
	Capability
	Maximum payload: 55,000 lb
	Up to 12 flights per year
	Satellite recovery, service, return

FIGURE 2.20
Space Shuttle Mechanical Subsystems. https://history.nasa.gov/diagrams/shuttle.htm.

The shuttle orbiter is by far the largest spacecraft ever launched into orbit. Although its main engines are used only during launch, a complex system of secondary thrusters and maneuvering engines gives the orbiter mobility and versatility in orbit. The payload bay can carry up to two satellites or the space laboratory into orbit, while the remote manipulator system is used for satellite deployment or retrieval and for orbital construction tasks. NASA's STS, consists of four shuttle orbiters, Atlantis, Discovery, Columbia, and Endeavor. The shuttles can carry satellites, experiments, and space station components into orbit and return them to Earth. The remote manipulator arm can handle objects that on Earth would weigh more than 30 tons. It also serves as a highly maneuverable platform for moving around space-walking astronauts. Space shuttle abort and normal mission profiles are shown in Figure 2.21. As was mentioned above, during the launch, the SRBs and the orbiter's three main engines fire simultaneously for more than 2 min. Then the SRBs are jettisoned while the shuttle engines continue firing for another 6 min or more. When they stop, the big external fuel tank is jettisoned. In the event of an emergency, during launch, pilots have several abort options depending on how high they are, including Abort-to-Orbit, in which a safe, but lower, orbit can still be achieved; Abort-Once-Around, in which the

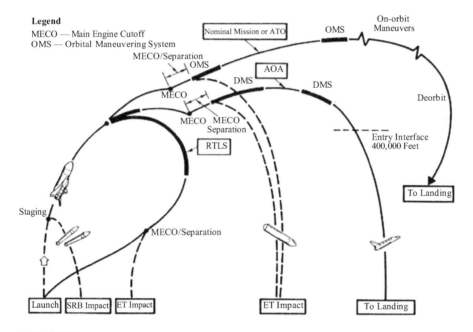

FIGURE 2.21
Abort and normal mission profiles. (Space Shuttle Detailed Block Program, http://www.google.com/search?hl=en&client=firefoxa&rls=org.mozilla:enUS:official&channel=np&q=space+shuttle+detailed+block+diagram&bav=on.2,or.r_gc.r_pw.&um=1&ie=UTF8&tbm=isch&source=og&sa=N&tab=wi&biw=1366&bih=624.[44])

shuttle executes one orbit and returns to Earth; Transoceanic Abort Landing, in which the shuttle lands at one of several sites in Europe or Africa; and Return-to-Launch Site, the most desperate option.

The ESA developed the Spacelab laboratory module, which could be installed in the cargo bay in different configurations, depending on the research mission. Astronauts enter the module through a tunnel to conduct experiments in life sciences, materials processing, astronomy, and Earth science. Different Spacelab configurations flew on 25 shuttle missions until the International Space Station was put into orbit. Astronauts inside the space shuttle orbiters are confined to a volume roughly the size of an average living room with up to seven crew members flying missions that can range from 5 to 18 days. The electricity for the orbiter is generated by fuel cells that combine stored hydrogen and oxygen. The chemical reaction produces electricity and, as a by-product, enough water for the crew. Prior to a shuttle launch, the orbiter must be prepared and the crew has to be trained.

The vehicle preparation takes place at the John F. Kennedy Space Center, Florida, over the course of 3–6 months—beginning with the orbiter's return from its previous mission. Then the refurbishment, mating to the external tank, and solid rocket boosters in the enormous vehicle assembly building (VAB) and integration of the next mission's payload follow.

Crew training can take several years for a complicated mission,[44] most of the crews' training takes place at Johnson Space Center in Houston, Texas.

One challenging task is docking in space. A typical space station docking begins with the orbiter approaching the station, usually from below. The shuttle is equipped with a radar navigation system that gives distances to the approaching target. Computers control the gradual approach, and the manual phase of the docking begins when the orbiter is roughly a half mile from the station. The pilot controls the orbiter from the craft flight deck while communicating with the station commander and the flight directors back on Earth. Beginning at about 30 ft from the station, the orbiter approaches at a rate of roughly 0.1 ft/s until it connects with the station's docking device. Following the fourth orbited test flight of Colombia, the space shuttle entered its operation on July 4, 1982. The shuttle formally entered service with a variety of satellites-related missions. The fifth flight of Columbia, STS-5, in November 1982 was the first to carry a commercial satellite on board—a mission that was seen as vital to establishing the shuttle as a commercial launch vehicle. In addition to the flight crew, STS-5 became the first mission to carry *mission specialists*—highly trained nonpilot astronauts.

The payload of two communication satellites deployed perfectly, and the only major problem on the mission was spacesuit flaw that prevented the first spacewalk of the shuttle program. The first full-scale science mission was Spacelab 1, carried aboard Columbia for the STS-9 launch in November 1983. By the end of 1985, the space shuttle program seemed to be getting into gear, with shorter intervals between flights and faster turnaround times for individual spacecraft. But then disaster struck. The launch of Challenger's

STS-51L mission on January 28, 1986, was the focus of unusual attention. It was the 25th flight, carrying a notable passenger—Christa McAuliffe from New Hampshire, who had been selected to become the first teacher in space and would be delivering lessons over a live television link from the shuttle to schools across the Unites States. But 73 s after launch, Challenger exploded in midflight, instantly killing all seven astronauts on board, showing debris across the Atlantic Ocean. Later, investigations showed that the night before launch, frost had formed on the shuttle, and a rubber O-ring seal had become brittle and unable to fill the joint properly under the stress of launch. Endeavor replaced Challenger after 32 months of spaceplane being grounded. The shuttle program resumed to launches of TDRS data satellite network in late 1988 and continued with launching classified, defense-related satellites. Spacelab missions resumed in 1990 with the ASTRO-1 astronomy mission. Challenger's replacement, Endeavour, finally took to the skies in May 1992 with the STS-49 mission. Spacelab was built by the ESA to fit into the shuttle cargo bay. The Spacelab laboratory module was first flown in 1983, as was previously mentioned, and helped to address NASA's lack of a permanent space station. The development for such a module was agreed upon in 1973 between NASA and ESA. The laboratory carried a wide range of experiments investigating everything from physics and materials science to space biology and astronomy, while its overall aim was to demonstrate the feasibility of carrying out such research in orbit. The Endeavour's STS-99 in 2000 carried radar that mapped the elevation of 80% of the Earth's surface. A blood sample taken during the first Spacelab mission revealed that weightlessness affects the production of red blood cells. The weightless environment of the shuttle allows physicists to see how materials behave away from the influences of gravity. NASA had decided to terminate the space shuttle program in 2011, counting on Russia to deliver payloads to the International Space Station (ISS). NASA's final mission finally launched on February 24, 2011. In a spectacular afternoon launch, the space shuttle Discovery blasted off on one final mission on February 24 to cap off its prolific 27 years of spaceflight.

Amid clear skies and warm temperatures, Discovery lifted off at 4:53 p.m. EST (2153 GMT) from Launch Pad 39A at NASA's Kennedy Space Center. "It was kind of an exciting last few minutes of this countdown. Several of us have been around for many, many countdowns, and this was one for the record books," said Mike Leinbach, NASA's shuttle launch director. "We were a couple seconds away from losing the window." The launch came as a relief after a last-minute glitch arose in a range-safety computer, threatening to prevent the liftoff. NASA cannot launch a shuttle unless the air space over Cape Canaveral is cleared by range safety officials. The issue was resolved within minutes of the launch, allowing Discovery to fly after all.

Recently, a privately owned U.S. company put a spacecraft into orbit and brought it back in a ground-breaking test flight NASA hopes will lead to supply runs to the ISS after the space shuttles are retired in 2011.[45] The NASA-backed mission was designed to try out a new system for delivering cargo and possibly crew someday to the orbital outpost. This was the first time a private company launched and returned a capsule from orbit. Space Exploration Technologies' Falcon 9 rocket lifted off at 10:43 a.m. from Cape Canaveral, Florida, and after two orbits of Earth, it parachuted to a splashdown in the Pacific Ocean more than 6 h later, on December 8, 2010. The company intends to upgrade its capsule with launch escape systems and hopes it will serve as a taxi for astronauts and other people wanting rides to the Station.

2.3.3.2 Buran—The Soviet Shuttle

For Soviets to have a similar system to that of the U.S. Space Shuttle to maintain parity was important, and the Soviets' decision to build and launch a space transportation system was announced in 1974, which at the time seemed quite logical. The spacecraft ended up looking similar to that of the United States, but its operating principle and launch hardware were quite different. Rather than build large solid rocket boosters, the design team decided on an entirely liquid-fueled rocket system, ultimately called the Energia, Soviets' Reusable Space System (MKS is its Russian acronym) (Russian: Энергия, *Energiya*, "Energy"). The orbiter known as Buran was designed so that it could carry an extra 5 tons of payload compared to the Shuttle, and Energia could function as an independent launcher, carrying payloads other than the orbiter. Due to some financial difficulties and for cost-cutting purposes, although the Buran successfully orbited the Earth twice and executed a perfect landing in 1988, the program was formally cancelled in 1993.

2.3.4 Russian Mir Space Station

Less than a month after the American space program had suffered the loss of Challenger, the Soviet Union launched the space station Mir. The development work on a complex modular space station began in the mid-1970s by NPO Energia (Korolev, Moscow area). The development progressed throughout the early 1980s but was sidetracked by other spacecraft programs, such as the progress Cargo Ferry and Buran. It only took priority once this Bureau was given a launch deadline of spring 1986, to coincide with the 27th Communist Party Congress. Mir was intended to be the first space station with more-or-less continuous occupancy. Mir, the world's first modular space station, used elements of the earliest civilian (DOS, Long-Duration Orbital Station; "DOS" in its Russian language abbreviation) Salyut stations, with additional laboratories and modules. By the end of 1990, Mir was approaching a decade of continuous operation during which a series of accidents threw the station's future into doubt. The most dangerous accident was the fire of February 1997.

After the Russian space station moved into its second decade, the Mir became notorious as an accident-prone spacecraft, even as it remained unparalleled in continuous service. A 15-min fire in an oxygen-generating device imperiled the station in February 1997. Failures of the Elektron electrolysis oxygen-generating units and problems with altitude and environmental controls often seemed to alternate with computer malfunctions and power outages.

The June 1997 collision with the Progress supply vehicle breached the integrity of the Spektr's hull and rendered that module uninhabitable. But Mir remained, and its space explorers endured. Over its lifetime, the space station hosted 125 cosmonauts and astronauts from 12 different nations. It supported 17 space expeditions, including 28 long-term crews.

2.3.5 The International Space Station

In 1984, President Reagan announced that the United States was at last going to build a permanent space station. Soon with the politically loaded name "Freedom," Europeans, Japanese, and Canadian space programs, joined the project, agreeing to provide their own laboratory modules and other elements. As the relationship between the United States and Russia improved, it led to a new way forward. The Russians had space station experience that could help NASA make its station a reality. In 1993, officials from the U.S. and Russian space agencies met to agree on a joint enterprise; the resulting station would be a hybrid of elements from Freedom and Russia's own stalled Mir 2. At first called Space Station Alpha, before long it became simply the ISS. The first element put into orbit was the Russian-built functional cargo block.

The module known as Zarya had a dual purpose: generating power and providing propulsion and acting as the station's heart during the early stages of construction. As the ISS grew larger and these functions moved elsewhere, it would become a storage facility.

The design of ISS has since changed with international contributions from Europe, Japan, and Canada. Table 2.1 shows the ISS characteristics, and Figure 2.22 shows the ISS configuration. The major components of the ISS are shown in Figure 2.23.

Since the first crew arrived in October 2000, the ISS has been continuously manned. As well as ongoing construction, the astronauts aboard work on a wide variety of scientific experiments. The following summarizes some of these experiments that were carried out on ISS, hosting state-of-the-art scientific facilities that support fundamental and applied research across the range of physical and biological sciences. From expedition 0–15, 138 experiments have been carried out on the ISS, supporting research for hundreds of ground-based investigators from the United States and international partners. Figure 2.23 also shows the distribution of experiments by discipline. Today, major research outfitting has grown to include 18 racks and facilities within the laboratory space as shown in Figure 2.24.

The focus of NASA's ISS research has changed strategically to support the revision for space exploration. While still including some fundamental

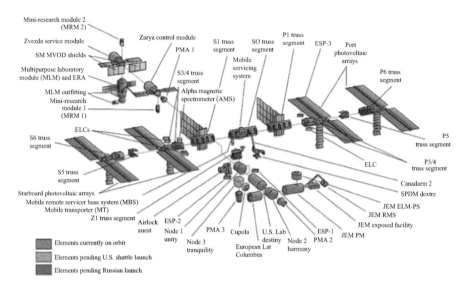

FIGURE 2.22
International Space Station (ISS) configuration.

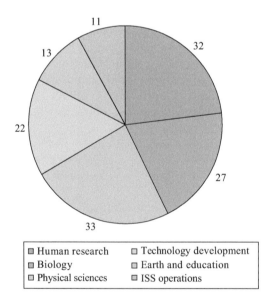

FIGURE 2.23
Distribution by discipline of the NASA International Space Station (ISS) experiments. (International Space Station Science research accomplishments during the assembly years: An analysis of results from 2000 to 2008. Cynthia A. Evans and Julie A. Robinson, Office of the International Space Station Program Scientist, NASA Johnson Space Center, Houston, Texas; Judy Tate-Brown, Tracy Thumm, and Jessica Crespo-Richey, Engineering and Science Contract Group, Houston, Texas; David Baumann and Jennifer Rhatigan, NASA Johnson Space Center, Houston, Texas, June 2009.[46])

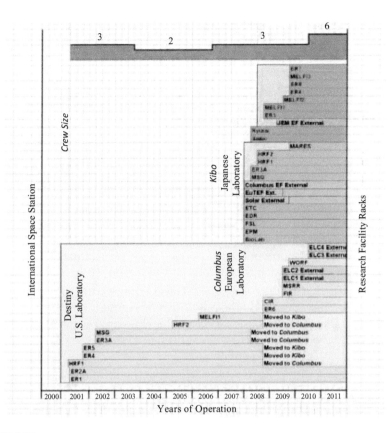

FIGURE 2.24
Eighteen racks and facilities within the laboratory space. (International Space Station Science research accomplishments during the assembly years: An analysis of results from 2000 to 2008. Cynthia A. Evans and Julie A. Robinson, Office of the International Space Station Program Scientist, NASA Johnson Space Center, Houston, Texas; Judy Tate-Brown, Tracy Thumm, and Jessica Crespo-Richey, Engineering and Science Contract Group, Houston, Texas; David Baumann and Jennifer Rhatigan, NASA Johnson Space Center, Houston, Texas, June 2009.[46])

research in microgravity, the emphasis has shifted to programs targeted at developing and testing new exploration technologies and reducing the risks to human explorers on missions to the Moon, Mars, and beyond. During the 2008 calendar year, the laboratory space and research facilities were tripled with the addition of the ESA Columbus and Japanese Aerospace Exploration Agency's Kibo scientific modules joining NASA's Destiny Laboratory.

By scientific theme, the investigations are categorized as follows[46]:

- *Technology development*—Studies and tests of new technologies for use in future exploration missions. Areas of emphasis included spacecraft materials and systems, and characterization and control of the microgravity environment on ISS.

- *Physical sciences*—Studies of physics and chemistry in microgravity. Areas of emphasis include materials sciences experiments including physical properties and phase transitions in polymers and colloids, fluid physics, and crystal growth experiments.
- *Biological sciences*—Studies of biology using microgravity conditions to gain insight into the effect of the space environment on living organisms. Areas of emphasis included cellular biology, biotechnology, and plant biology.
- *Human research for exploration*—Human medical research to develop the knowledge that is needed to send humans on exploration missions beyond Earth's orbit. These studies focused on the effects of living in space on human health and countermeasures to reduce health risks that will be incurred by living in space in the future. Areas of emphasis include physiological studies related to the effect of microgravity on bone and muscle, other physiological effects of space flight, psychosocial studies, and radiation studies.
- *Observing the Earth and education*—These activities and investigations all allowed students and the public to connect with the ISS missions; inspired students to excel in science, technology, engineering, and math; and shared the astronauts' unique view of the Earth system with scientists and the public.
- *Results from ISS operations*—In addition to the formal, peer-reviewed scientific research and experiments, the ISS supports a large body of research using data from ISS operations, including routine medical monitoring of the crew and data that are collected on the ISS environment, both inside and outside of the ISS.

The most recent Expedition to ISS that is completed is Expedition 52 which started on June 2, 2017 with the unlocking of Soyuz MS-03 with crew members.

The research activities carried out onboard ISS have resulted in hundreds of publications in variety of disciplines. Figure 2.25A and B show the publications of the ISS scientific results.[46]

2.3.5.1 Next Generation Space Stations

Since November 2000, people have continuously lived and worked in space on the ISS, and have made significant contributions to improving life on Earth and enabling future spaceflight activities.[47] However, the ISS is scheduled for destruction in either 2020 or 2024, although there is reason to believe it can last until at least 2028. If that time comes with no replacement, America's and humanity's hard-won foothold in LEO will be lost. Without adequate planning, the end of the ISS program will result in the loss of a host of valuable capabilities and activities that promote commerce, science,

(A) **Research Discipline of ISS Investigations By Partner Agency:**
Expeditions 0-40
December 1998 - September 2014

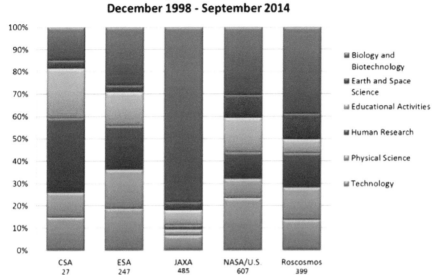

(B) **ISS Results Publications through October 2015**

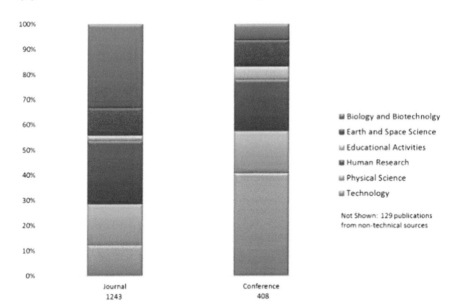

FIGURE 2.25
(A) Distribution by discipline of NASA's ISS experiments for Expeditions 0–40. Graph courtesy of JSC/NASA. (B) A compilation of the number of publications resulting from research aboard the ISS. Each discipline has produced a steady stream of published results. https://www.nasa.gov/sites/default/files/atoms/files/iss_technical_publication_030116.pdf.

space operations, and space settlement. These include human-tended materials research, biological research, physics research, robotics, satellite launch, Earth observation, and astronomy, all conducted by commercial firms, academic institutions, and governments from around the world. We will also lose the only space hotel that has ever hosted a paying guest and a valuable example of extensive peaceful international cooperation. NSS (The National Space Society) suggests the followings:

- Variable gravity research, including life support development targeted toward the Moon and Mars, and the effects of lunar and Martian gravity levels on plants and animals.
- Microgravity research, including materials and biological research, with large-scale manufacturing spun off to additional commercial stations as research is productized.
- Space operations (fuel depots, cargo/crew transfer, lunar transit, and pair/refurbishment facilities), perhaps in high orbit. Regular access to LEO, both to bring new experiments and materials to the ISS or future space stations, and to return experimental outputs and products to the Earth, is essential to both the scientific and commercial utilization of LEO.

Since the retirement of the Space Shuttle, it is only with the advent of the Commercial Resupply Services (CRS) program that regular access to and from the ISS on U.S. vehicles is possible, and that is essential for full utilization of the ISS National Laboratory. Hence, when we look toward a post-ISS transition to a number of commercial stations, continued NASA support for routine access to LEO via CRS is a fundamental requirement. The recent announcement that in addition to the two incumbents SpaceX and Orbital Sciences, Boeing, Sierra Nevada, and Lockheed Martin say they will bid on CRS-2.

2.3.6 Nanosatellites and Microsatellites

In previous sections, the different types of small satellites were briefly touched upon. Here we look at this area of satellites in more details and will discuss the technology and potentials of Cubesats.

The following Figures 2.26–2.28 show the excising and projected nano/microsatellites through 2020.

Cubesats emerged in 1999 as a tool for university STEM education. As their popularity grew, an ecosystem developed around this form factor and is a key driver of today's expanding nanosatellite market. One quarter of nano and microsatellites launched between 2010 and 2014 were 1 Unit (1U) cubesats (10 cm^3), and the operator pool has grown from universities to commercial, government, and military players. But as the small satellite market continues to grow, will 1U cubesats persist as a standard or be considered inadequate in favor of larger/more powerful satellites?

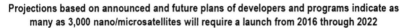

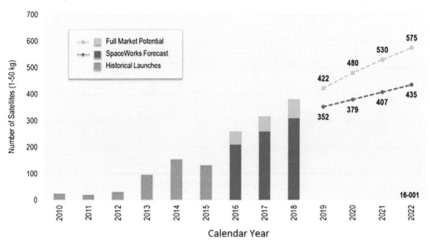

FIGURE 2.26
The 2016 Full Market Potential dataset based on the SpaceWorks Forecast dataset. http://
aacmicrotec.com/spacecraft-bus-and-subsystems/.

Large Program Breakdown for Announced Future Satellites

Name of Program/ Satellite Constellation	Timeframe	Organization	Country	Mass (kg)	Launched to Date	Total Planned
NSF Geospace & Atmospheric CubeSat	2010-2015	NSF	USA	1-3	7	13
NASA EDSN	2013-2014	NASA ARC	USA	3	0	8
NASA CubeSat Launch Initiative	2011-2017	NASA	USA	1-12	24	115
SeeMe Payloads	2016	DARPA	USA	12	0	6
QB50	2015	Von Karman Institute / Various	Various	2	0	52
HUMSAT	2013-2014	University of Vigo / Various	Various	1	0	9

Existing large programs will comprise only 25% of future nano/microsatellites (compared to 65% in 2013) due to worldwide growth in the civil and commercial sectors

FIGURE 2.27
Nano/micro satellite future programs. Nano/microsatellite Market Assessment. https://www.
emaze.com/@AFZTOZW/Nanosatellite-Industry-Overview-updated-022014.pptx.

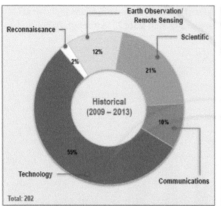
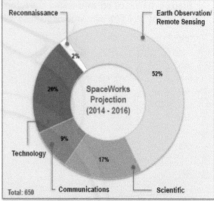

A smaller proportion of technology development/demonstration nano/microsatellites will be built in the next few years (20% vs. 55% from 2009 to 2013)

FIGURE 2.28
Nano/Microsatellite trend by Purpose (1–50 kg). https://www.slideshare.net/prateepbasu/nano-and-microsatellite-market-assessment2014.

Evaluating Utility of the 1U:

For small sat operators, mission cost, ability to leverage COTS components, launch opportunities, and ultimately payload capabilities drive form factor decisions. With larger size and greater payload capabilities generally come increases in cost, timeline, and mission complexity. Where does the optimal balance lay?

Figure 2.29 shows the comparison in size between different cubsats.[48]

- **Too Large**

  Few potential users deem 1U satellites too large. That said, pico and femtosatellites, including the standardized PocketQube form factor,

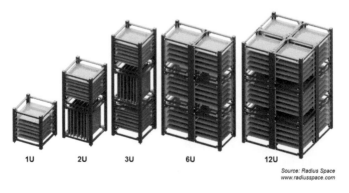

Source: Radius Space
www.radiusspace.com

FIGURE 2.29
Cubesat size comparison.

have been pursued by educational users as well as amateur radio and space aficionados. The principal draw thus far is obtaining space experience within a more manageable scope and decreased manufacturing and launch price compared to a full cubesat, and this will lead a small population of players previously in the 1U target market to turn to smaller platforms moving forward.

- **Too Small**

  The single largest concern voiced about the 1U form factor is its limited payload accommodations (mass, volume, and power). Despite prior commercial and government missions using the 1U form factor, these users are expected to opt for larger platforms in the future. Only one announced commercial venture leverages a 1U design, and the U.S. National Science Foundation recently stated that 1U cubesats are simply too small to accomplish meaningful science. The fastest growing nano and microsatellite operator community falls into this category.

- **Just Right**

  The original target user of the 1U cubesat—universities—are still very much the key market. 1Us provide the right balance of low costs, limited project complexity and timeline, and hands-on experience for a student team and (low) university resources. As universities gain experience and establish funding streams that enable them to pursue more ambitious larger missions, novice universities and even high schools are emerging to take their place in the 1U operator segment. Despite plans for the commercial Outernet constellation using 200 1U cubesats, NSR (Northern Sky Research) does not anticipate full deployment of this system or ongoing use of this platform for commercial purposes.

NSR's Nano and Microsatellite Markets, 2nd Edition found that the 1–3 kg nanosatellite segment—largely composed of 1U cubesats—will grow at a 3% CAGR over the next decade. Just over 500 such satellites are expected to launch by 2024. While this represents positive growth, it also belies the reduction in market share of this mass segment: a fall from 38% in 2010 to 12% in 2024 as shown in Figure 2.30. Rather, the single-most active small satellite segment is expected to be 3–10 kg platforms. Larger 10–100 kg platforms are the most rapidly growing, at a 14% CAGR.

A Preference for Larger Form Factors

In 2014, the 1U cubesat's larger version, 3U, became the most prevalent form factor in this market. This was primarily tied to deployment of the Planet Labs constellation, but the platform is also integrated in plans for other commercial systems as well as a steady stream of government and university science applications. While constellations using 3U platforms will sustain activity in this market through the next decade, NSR expects the next wave of growth to emerge with 6U and 12U cubesats. These larger cubesats strike

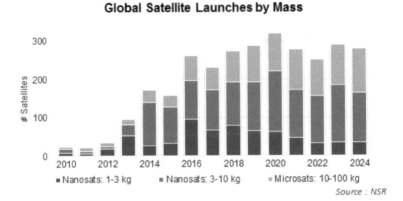

FIGURE 2.30
Global satellite launches by mass from 2010 to 2024.

a balance between enabling far more capable payloads, but limiting manufacturing and launch costs and leveraging the benefits of standardization.

Bottom Line
1U cubesats are a driving force behind today's resurgent interest in small satellites. Simplicity and the concepts of standardization and COTS components championed through 1U cubesats opened the satellite market to diverse users, and these same ideas are now being successfully applied to larger form factors, both multi-U cubesat and otherwise. Yet, the demand for increased payload capacity to support diverse applications and more capable instruments is leading the market to larger form factors, and leaving 1U platforms to fulfill their original purpose as an educational tool. But in this dynamic and rapidly evolving market, the ongoing march of technology miniaturization could revive 1U utilization in the 2020s and beyond. (Source: www.nsr. com/news-resources/the-bottom-line/mass-challenge-for-cubesats/).

Cube Sat Launches Growing Dramatically as seen below in Figure 2.31 over 15 years.

As it could be observed from the following graph, Figure 2.32, and the data provided before, Earth Observation is the fastest growing application of Cubesats.

2.3.6.1 Growing Deployment of Small Satellites

The development of small satellites[49] for Earth imaging and establishing space-based Internet networks from LEO is now possible because satellite components have been miniaturized and standardized. Groups of small satellites are referred to as constellations; Planet Labs, for example, has a constellation of 36 small satellites in orbit, with customers paying for the images it can capture at less distance from Earth than is possible with larger

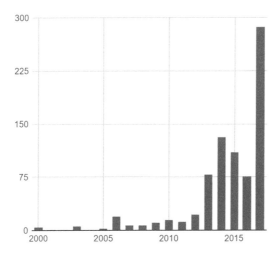

FIGURE 2.31
CubeSats Launched Each Year (2000-Present, 798 Spacecraft).

Source: **https://sites.google.com/a/slu.edu/swartwout/home/cubesat-database.**

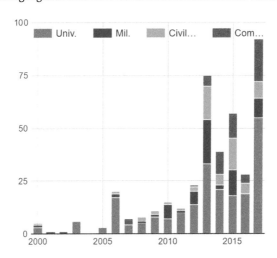

FIGURE 2.32
CubeSata by Mission Type (2000-Present, 403 Spacecraft, no Constellations). https://sites.google.com/a/slu.edu/swartwout/home/cubesat-database.

satellites in higher orbits. Some observers argue that demand for data may be driving the market for small satellites, as much as the new technologies.[50]

Small satellites allow access to space by researchers, companies, and governments that cannot afford larger spacecraft. Because small satellites can travel as a secondary payload on many launch vehicles, launch costs may be only a few million dollars per satellite, although multiple small satellites may be required for many purposes.[48]

The small satellite market is addressed by a number of startup firms that hope to succeed in providing broadband, remote imaging, or communication services, such as Firefly Space Systems in Texas, Rocket Lab in California, and OneWeb in Virginia. While small satellites are normally launched on rockets with other, larger payloads, Virgin Galactic is proposing a new type of launcher: a rocket attached to the wing of a modified commercial 747 jet will launch a payload into orbit when the plane reaches an altitude of 35,000 ft. Virgin Galactic believes this form of launch vehicle, using smaller rockets, will significantly reduce the cost of putting small satellites into orbit.

NASA is also utilizing small so-called CubeSats to address scientific questions and broaden the involvement of students and researchers. In 2015, 42 small satellites, primarily with commercial functions, were launched as cargo to the ISS for later deployment from there. NASA has announced that it will help develop new CubeSat technologies and will launch six small Earth-observing satellite missions.

2.3.7 International Space Activities

To get a sense of global market for space activities, we will look at some data for the current status as well as the future forecast up to 2022. Figure 2.33 shows that the global expenditure has increased from $30 billion in 1994 to over $70 billion at the present time and will reach over $80 billion by 2022. Figure 2.33 also shows the U.S. space expenditure in civil and defense segments in which the civil shows an steady state and small increase where defense expenditure is reducing.

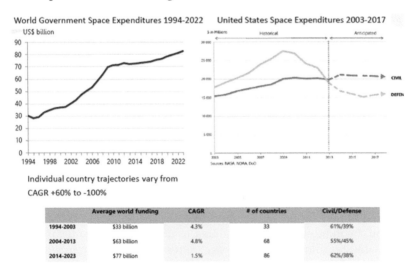

	Average world funding	CAGR	# of countries	Civil/Defense
1994-2003	$33 billion	4.3%	33	61%/39%
2004-2013	$63 billion	4.8%	68	55%/45%
2014-2023	$77 billion	1.5%	86	62%/38%

FIGURE 2.33
Growing Global Public Expenditures.

Source: **Euroconsult** *Government Space Programs: Forecast and Benchmarks 2014.*

Figure 2.34 shows the number of non-defense launches of satellites for several countries from the early 1961–2014.

Growing Global Public Expenditures

Figure 2.35 below shows entrepreneurial activity growth in small satellites and data analytics for several countries in which the United States shows its dominion.

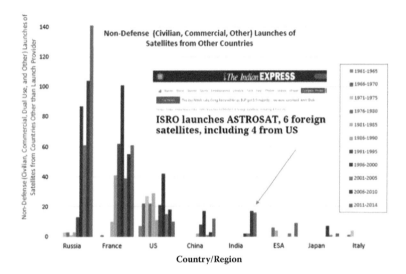

FIGURE 2.34
Non-Defense Satellite Launches by several Contries.

Source: **STPI generated using data from McDowell 2015.**

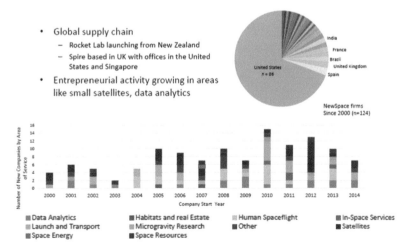

FIGURE 2.35
United States entrepreneurial activity growth in small satellites.

Figure 2.36 below shows the countries involved in space activates for several regions worldwide. The space activities are listed for several application areas as well as small satellites and launch activities. The colors are indicative of the stages that the countries are engaged in space activity.

Figure 2.37 shows the number of total launches in 2014 worldwide.

		Earth Observation	Communication Satellite Services	Space S&T and Exploration	Launch and Access to Space	Position, Navigation, and Timing (PNT)	Human Space Flight	Space Situational Awareness (SSA)	Small Satellites
Africa and Middle East	Algeria								
	Iran								
	Israel								
	Nigeria								
	South Africa								
	Turkey								
	UAE								
	Angola								
	Congo								
	Egypt								
	Gabon								
	Ghana								
	Kenya								
	Morocco								
	Saudi Arabia								
	Tunisia								
Asia	Australia								
	China								
	India								
	Indonesia								
	Japan								
	Malaysia								
	South Korea								
	Taiwan								
	Thailand								
	Vietnam								
	Bangladesh								
	Laos								
	North Korea								
	Pakistan								
	Singapore								

		Earth Observation	Communication Satellite Services	Space S&T and Exploration	Launch and Access to Space	Position, Navigation, and Timing (PNT)	Human Space Flight	Space Situational Awareness (SSA)	Small Satellites
Europe, Western	Austria								
	Belgium								
	Czech Republic								
	Denmark								
	ESA								
	Eumetsat								
	EU								
	Finland								
	France								
	Germany								
	Italy								
	Luxembourg								
	Netherlands								
	Norway								
	Poland								
	Spain								
	Sweden								
	Switzerland								
	UK								
Europe, Eastern	Bulgaria								
	Cyprus								
	Estonia								
	Greece								
	Hungary								
	Ireland								
	Latvia								
	Lithuania								
	Portugal								
	Romania								
	Slovakia								
	Slovenia								

FIGURE 2.36
Country Involvement in Space.

Source: **STPI analysis using data from Euroconsult (2014a).**

(Continued)

		Earth Observation	Communication Satellite Services	Space S&T and Exploration	Launch and Access to Space	Position, Navigation, and Timing (PNT)	Human Space Flight	Space Situational Awareness (SSA)	Small Satellites
Latin America	Argentina								
	Bolivia								
	Brazil								
	Mexico								
	Venezuela								
	Chile								
	Colombia								
	Ecuador								
	Nicaragua								
	Peru								
North America	Canada								
	United States								
Russia and CIS	Belarus								
	Kazakhstan								
	Russia								
	Ukraine								
	Armenia								
	Azerbaijan								
	Mongolia								
	Turkmenistan								

White	Orange	Turquoise	Blue
No interest or activity	Interest and minimal development	Operating or near-operating capability with international partnerships	Fully fledged and independent capability

FIGURE 2.36 (CONTINUED)
Country Involvement in Space.

Source: STPI analysis using data from Euroconsult (2014a).

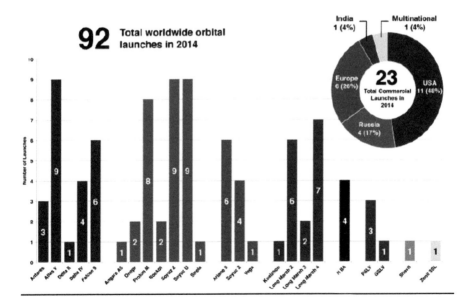

FIGURE 2.37
Number of total (and Commercial) launches, 2014.

Source: FAA 2015, http://space.taurigroup.com/reports/FAA_Annual_Compendium_2014.pdf.

Space activities spread to other countries in the early 21st century, and in what follows, some of these countries will be mentioned, although there are more emerging countries in this field.

2.3.7.1 Russian Space Activities

Russia's space program is organized around the federal space agency, Roscosmos, as well as a branch of the armed forces dedicated to all military satellites and launch facilities. The agency focuses on practical applications, such as communications, remote sensing, and navigation. Its main priorities include reconstituting aging satellite fleets, space sciences, and human spaceflight. Russia continues to be a world leader in commercial launch services, and continues to make investments in launch infrastructure and systems. Russia also continues to allocate funds to support the ISS, including crew and cargo transportation. With the retirement of the U.S. Space Shuttle, Russia will be solely responsible for the transport of crew to and from the ISS for the foreseeable future. The oldest and most widely participated body is the UNCOPOUS that provides a platform for addressing broad policy and legal problems arising from the use and exploration of outer space. The Group on Earth Observations (GEO) and the International Space Exploration Coordination Group are also classical examples of the increasing willingness of nations to cooperate at a multinational level.

The following Figure 2.38[51] shows the Russian space programs from 2010 to 2040 in several key areas.

In 1957, the Soviet Union put the first satellite, Sputnik, into orbit. In 1961, it launched the first human, Yurii Gagarin, into space. From that time until its dissolution in 1991, the USSR maintained a robust space program, often following lines of development very different from its one major competitor, the United States. However fast the political and economic landscape may be changing in Russia, the speed with which the space program can change and the directions it can take are constrained by how it developed during the Soviet era. Since the USSR era still has large residues in the Russian space activities, in the following we will take a quick look at that era.

Russia's civilian space program is still using equipment and material manufactured and stored before the dissolution of the USSR, such as stockpiles of Proton rockets, satellites, and the Mir Space Station. Some of the impetus for the high-level production was a desire to equal or surpass U.S. accomplishments in space. Figure 2.39, which shows the number of launches in the United States and in the USSR since 1957, not only demonstrates the productive capacity of the USSR's space industry, but also indicates the difference in design philosophies of the two countries. Where the United States built long-lived, technically sophisticated payloads, the USSR built much shorter-lived satellites that required more frequent replacement. The difference in design philosophy between the two countries goes back to the origins of their space programs. At the end of World War II, the German rocket scientists from

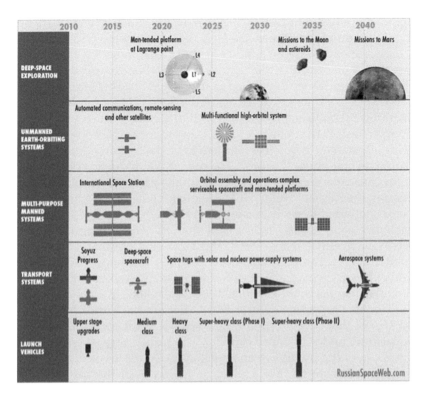

FIGURE 2.38
A timeline unveiled by the Russian space agency in 2013 outlining general direction of the
Russian space program.

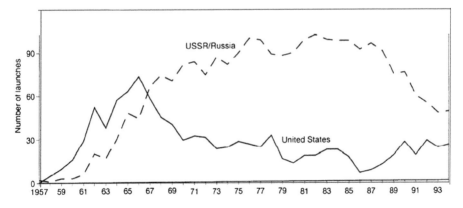

FIGURE 2.39
Successful United States long-lived and USSR short-lived launches. http://link.library.
missouri.edu/portal/Space-activities-of-the-United-States-CIS-and/_y_Zz-aHjzQ/.

Source: **Marcia S. Smith,** *Space Activities of the United States, CIS, and Other Launching
Countries/Organizations: 1957–1993,* **Congressional Research Service Issue Briefs,
Washington, DC, March 29, 1994.**

Peenemünde, who were responsible for the V-2 Rocket, were part of the spoils of war divided between the United States and the USSR Both countries used the experience and skills of these men to set up their ballistic-missile programs. Because the Soviet hydrogen bomb was so much larger and heavier than the one developed in the United States, it required a larger, more powerful rocket to carry it. In fact, the Soviet Union produced the first intercontinental ballistic missile (ICBM) in the world, which was known in the West as the R-7. The Soviets' expertise in producing rockets with large lift capacities then made it possible for them to be the first to produce launchers that could carry humans into orbit. The successful production of rockets with large lift capacities reduced incentives to make payloads compact and light. To this day, the Russian satellite is less sophisticated in its electronics (Soviet satellites continued to use vacuum tubes long after the West had switched to solid-state components) and heavier than its Western counterparts. The Soviets built satellites with a much shorter design life than was typical in the United States. The requirement to maintain these space systems led to the Soviet Union's remarkable (by Western standards) ability to replace damaged or obsolete satellites. For some types of satellite, the Soviet Union was able to launch a replacement in 24 to 48 h. Both in the United States and the Soviet Union, the space program was a symbol of the country's technological superiority and productive capacity. The United States kept its military program out of the public eye and created the NASA as a separate civilian space program, with its own budget, as the focus of the national civilian space effort.

The Soviet Union, on the other hand, never created separate civil and military space programs; the same budget supported both efforts.

Much of the same infrastructure, production organizations, design bureaus, and personnel were used to service both programs.

2.3.7.2 United States Space Activities

United States has been and continues to be in the forefront of space activates in the recent modern history. The space activities of the United States covers a wide spectrum of technologies that by itself deserves volumes. In this book, the reader is exposed to such technologies in different chapters and sections and in every topic that is being discussed.

The United States National Space Policy published in June 2010 summarizes these activities in its statement of Goals which are as follows:

Consistent with the principles set forth in this document*, the United States will pursue the following goals in its national space programs:

- Energize competitive domestic industries to participate in global markets and advance the development of: satellite manufacturing;

---

* https://www.nasa.gov/sites/default/files/national_space_policy_6-28-10.pdf.

satellite-based services; space launch; terrestrial applications; and increased entrepreneurship.

- Expand international cooperation on mutually beneficial space activities to: broaden and extend the benefits of space; further the peaceful use of space; and enhance collection and partnership in sharing of space-derived information.

- Strengthen stability in space through: domestic and international measures to promote safe and responsible operations in space; improved information collection and sharing for space object collision avoidance; protection of critical space systems and supporting infrastructures, with special attention to the critical interdependence of space and information systems; and strengthening measures to mitigate orbital debris.

- Increase assurance and resilience of mission-essential functions enabled by commercial, civil, scientific, and national security spacecraft and supporting infrastructure against disruption, degradation, and destruction, whether from environmental, mechanical, electronic, or hostile causes.

- Pursue human and robotic initiatives to develop innovative technologies, foster new industries, strengthen international partnerships, inspire our Nation and the world, increase humanity's understanding of the Earth, enhance scientific discovery, and explore our solar system and the universe beyond.

- Improve space-based Earth and solar observation capabilities needed to conduct science, forecast terrestrial and near-Earth space weather, monitor climate and global change, manage natural resources, and support disaster response and recovery.

2.3.7.3 Chinese Space Activities

Chinese space-related government organizations include the China National Space Administration (CNSA), an internal structure of the State Administration for Science,[52] Technology and Industry for National Defense (SASTIND) and the state-owned China Aerospace Science and Technology Corporation (CASC). The manned spaceflight program is operated by the Chinese Manned Space Engineering Office (CMSEO), which is within the General Armaments Department (GAD) of the People's Liberation Army (PLA). Funding for space is part of military budgets, which are only partially disclosed. China's goals include facilitating economic development, ensuring self-reliance, promoting national prestige, and projecting power. Programs focus on manned spaceflight and space applications, such as remote sensing, communications, navigation, and space science and technology. In parallel, China has developed capabilities to limit or prevent the use of space-based assets by potential adversaries during times of crisis or conflict. The country launched its first astronaut (taikonaut) into Earth's orbit in 2003, conducted an ASAT test in 2007 on its aging weather satellite and performed its first spacewalk in 2008. In 2010, China successfully deployed its second lunar probe, Chang'e-2, to scout possible landing sites for an unmanned Chang'e-3 mission.

China embarked on an ambitious project to equal the achievements of the 20th-century superpowers, sending men into orbit and perhaps beyond. Although China planned a manned space-flight program as early as the 1960, the Chinese government approved a new manned space program, initially known as Project 921, in 1992. The newly formed CNSA benefited from an agreement signed with Russia in 1994, which gave them access to Soyuz Capsules, blueprints, and Russian expertise. A new Chinese vehicle called *Shenzhou* (meaning "divine vessel") is completely Chinese in design and manufacture, despite Russian assistance. Shenzhou has three separate elements: an orbital module, a reentry module, and a service module. It is significantly larger than Russian's Soyuz and is fitted with two sets for solar arrays—one pair on the service module, the other on the orbital module.

Shenzhou 1, an unmanned development test, was launched in November 1999 by a CZ-2F Long March rocket. It orbited Earth 14 times, and its central module landed safety in Inner Mongolia. On October 15, 2003, Shenzhou 5 was launched with a Chinese spacefarer (or *taikonaut*) on board. It orbited Earth 14 times before landing in Inner Mongolia as did the Shenzhou 1 Mission. The *taikonaut* remained in the reentry module throughout the flight, but the experiments on board the orbital module continued to function for 5 months after the craft had been abandoned in space. Shenzhou 6 was launched in 2005, and this time two *taikonauts* remained in orbit for almost 5 days. China's plans for the future are as follows:

Early 2010s—Space station

2024—Manned lunar landing

After 2040—Missions to Mars

China launched its first satellite in 1970, which was an updated equivalent of Russia's Sputnik 1, orbiting the Earth. The launch vehicle was Chang Zheng ("Long March," abbreviated as either CZ or LM design) which is still in use today. The success of the CZ-rocket allowed China to enter the commercial launch market in 1985. Early customers included the Communications Satellite Corporation (COMSAT) companies, Asiasat in Hong Kong and Optus in Australia, as well as organizations in Sweden and Pakistan. Much of the Iridium satellite phone network has also been launched from China. Since its first launch, China has put more than 50 satellites into orbit with a wide range of applications, including remote-sensing atmospheric studies, and a system of regional communication satellites also known as Chinasat.

China, announced on October 27, 2010, that it plans to launch the first part of a manned space station by 2016 and to complete a "relatively large" laboratory by around 2020. Having sent its first astronaut in 2003, China also plans to become the second country after the United States to land a man on the Moon by 2025. This is at times when the United States has canceled its manned lunar program (February 2010) and is grounded its space shuttle in 2011, relying entirely on Russia, at least through the first half of the decade, to take astronauts to the ISS.[53]

The Chinese Space Station (CSS) with the following specifications is underway. The modules of CSS and new types of launch vehicles as well as other related facilitates are under development.[54]

Chinese Space Station

Design specification Number of Modules: 3

Inclination: 42°~43°

Altitude: 340~450 km

Lifetime: ≥10 years

Crew members: 3

Maximum 6 crew members during rotation.

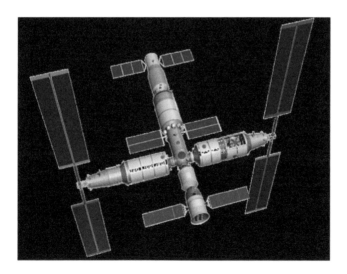

The Construction Plan for CSS is as follows:

- The Testing Core Module is scheduled to be launched in 2018; several manned spaceships and cargo spaceship will be launched to visit the Testing Core Module, conducting key technique tests such as on-orbit assembly, EVA, long-term manned flight, etc.

- The Experiment Module I and II will follow afterwards.
- The Space Station with three modules will be put into operation around 2022.

The experiments envisioned for the CSS include the followings as shown in Figure 2.40.

China is also developing a system to recover and to reuse space rockets.

This move will reduce costs of space program and reduces risk of debris from launches falling to Earth.

Space Science Experiments

The three modules of CSS will be featured with advanced technologies and equipped with multi-purpose facilities in international standards for space science.

Space life sciences and biotechnology
- Ecology Science Experiment Rack (ESER)
- Biotechnology Experiment Rack (BER)
- Science Glove-box and Refrigerator Rack (SGRR)

Microgravity fluid physics & combustion
- Fluids Physics Experiment Rack (FPER)
- Two-phase System Experiment Rack (TSER)
- Combustion Experiment Rack (CER)

Material science in space
- Material Furnace Experiment Rack (MFER)
- Container-less Material Experiment Rack (CMER)

Fundamental Physics in Microgravity
- Cold Atom Experiment Rack (CAER)
- High-precision Time-Frequency Rack (HTFR)

Multipurpose Facilities
- High Micro-gravity Level Rack (HMGR)
- Varying-Gravity Experiment Rack (VGER)
- Modularized Experiment Rack (RACK)

FIGURE 2.40
CSS space science experiments.

The plan is to recover parts of rockets used in space launches to bring down costs and make its space program more commercially competitive, according to researchers involved in the project.

The system would bring the rocket engine and booster safely back to the ground so they can be reused in future launches. Besides saving operational costs, the recovery would also reduce the threat of debris falling to the ground, the researchers said.

The recovery system is underdeveloped at the China Academy of Launch Vehicle Technology in Beijing.

It involves using a set of multiple parachutes, which are stored in the first stage of the rocket, which is released from the rest of the craft before it burns its way through the Earth's atmosphere.

An airbag inflates under the discarded part of the rocket, which cushions impact when it finally hits the ground.

China is planning to launch its most powerful rocket closing the gap with United States.

The technology that the Chinese will be using differs from the system used by the commercial company SpaceX on its Falcon 9 rockets. As the first stage of Falcon 9 falls back to Earth, its engines reignite when it reaches a speed of 3 km per second, slowing it down to reduce impact as it lands vertically on the ground.

In an article on the academy's website, Deng Xinyu, a researcher on the Chinese rocket recovery program, said that vertical landing involved many challenges and was extremely difficult to achieve.

SpaceX has performed a series of successful vertical landings which demonstrate the technology's feasibility, but Chinese researchers have rejected the approach.

The vertical landing system needs carrying extra fuel for landing, which meant rockets could only carry smaller payloads into space.

The Falcon rocket also uses nine small rocket engines to generate thrust, which reduces the launch vehicle's overall reliability and efficiency. "The mainstream trend of modern rocket development is to increase the thrust and reduce the number of rocket engine. That is also why China, as well as Airbus, Boeing and Lockheed Martin did not use the technology."

China's government has funded research into both approaches before concentrating efforts on the parachute system.

Another activity that the Chinese are undertaking is to develop space rockets to launch from planes.

A large-scale test experiment using the technology was carried out 2 years ago.

Bao Weimin, professor of aerospace technology at Peking University and a science advisor to the China Aerospace Science and Technology Corporation, told the state-run news agency Xinhua that the reuse of rockets was essential to cut costs.

"Compared to mainstream rockets overseas, Long March rockets in China have lower costs, but with the increase of Chinese space exploration, the costs must be trimmed further," he was quoted as saying.

China has yet to officially rule out the vertical landing technology to recover rockets, according to Xinhua. A final decision will be made before 2020, the report said.

The Chinese launch vehicle academy researchers have said on their website that the parachute-airbag system will be assisted by multiple sensors and a sophisticated flight control system to guide the descent. The final goal was to hit a small recovery zone as precisely as the vertical landing technology can achieve.

Figure 2.41 shows a snapshot of Chinese space activities from 1970 to 2025.

2.3.7.4 Japan's Space Activities

The National Space Development Agency of Japan (NASDA) was established in 1969 charged by the government with turning Japan into a major space power. NASDA was formed after the Institute of Space and Astronautical Science (ISAS) was funded in the 1950s. ISAS's concentration was on space research projects, such as astronomy satellites and planetary probes, while NASDA focused on the development of commercial launch vehicles, other satellite applications, and manned spaceflight.

In 2003, the Japan Aerospace Exploration Agency (JAXA) was formed by the mergers of ISAS, NASDA, and NAL, Japan's National Aerospace Laboratory. JAXA has plans for a probe to Venus and is involved in the Bepicolombo mission to Mercury, a joint venture with ESA (see Figures 2.42 and 2.43). JAXA's current launch vehicle is the H-IIA, a basic two-stage rocket, fueled by efficient liquid hydrogen and oxygen to which a variety of booster modules can be added if necessary, depending on the size of the payload and its destination. JAXA's roadmap is shown in Figure 2.42. The H-IIA was developed from NASDA's earlier H-II which used similar fuels and had similar flexibility. JAXA is now developing the more powerful H-IIB, based on H-IIA's power technology.

1970: China launches its first satellite.

2003: The first *taikonaut* (from *laikong*, Chinese word for "space") is sent into orbit.

2007: China shoots down its own weather satellite to test its ballistic missile capability.

2007: China's first unmanned lunar probe is launched.

2008: The first *taikonaut* space walk takes place.

2016: China plans to launch first part of space station.

2017: China aims to bring Moon rocket back to Earth.

2025: China plans to land first *taikonaut* on the Moon.

FIGURE 2.41
China's race to space.

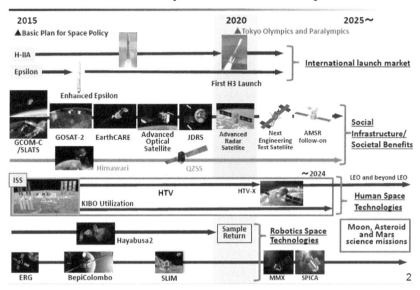

FIGURE 2.42
JAXA'a roadmap: 2020 and beyond.[55]

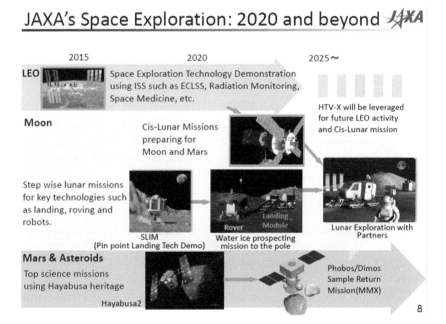

FIGURE 2.43
JAXA's space exploration: 2020 and beyond.[55]

The Ministry of Land, Infrastructure, Transport and Tourism is responsible for policy and planning implementation while the Ministry of Education, Culture, Sports, Science and Education (MEXT) is in charge of the Japanese space budget. Japan's Maritime Self-Defense Force (SDF) and the Japanese Defense Agency (now Ministry of Defense) have maintained cooperation with JAXA, with two of JAXA's first astronauts serving as SDF officials. However, the utilization of space for national security purposes has been severely restricted to quasi-commercial space assets through the 1969 Diet Resolution Concerning the Peaceful Utilization of Space.

The management structure of the Japanese space program, along with policy formulation and Japan's definition of what constitutes peaceful use of space underwent a significant change with the enactment of the Japan Basic Space Law. The Basic Space Law established a Strategic Headquarters for Space Policy. This is Japanese terminology for a new and higher level space governance structure. The administration of the Democratic Party of Japan elevated the importance of the Strategic Headquarters for Space Policy by announcing that its' Chair would be the Prime Minister with Acting Chairs being the Chief Cabinet Secretary and the Minister of Space Development. The Basic Space Law also provides that space will be used for the purpose of "improving and reinforcement of information gathering functions and enhancement of warning and surveillance activities in light of the international situation, especially the circumstances in North East Asia." This is indicative of a major shift in Japanese policy by eradicating the long-standing restrictions on the involvement of the Ministry of Defense in space activities.

The Basic Plan for Space Policy that was released by the Strategic Headquarters for Space Policy in June 2009 is a 5-year plan with a 10-year horizon. The Plan articulates the following "Six Basic Pillars for Japan's use and R&D of space":

1. **Ensure a rich, secure, and safe life:** Improve the national space-based infrastructure for more effective weather forecasting, telecommunications, food and energy management, positioning and navigation services, etc.

2. **Contribute to enhancement of security:** Strengthen the national information-gathering capability.

3. **Promote the utilization of space for diplomacy:** Provide access to satellite imagery to Asian neighbors for disaster prediction and management, climate change research, and other global environmental concerns.

4. **Create an energetic future by promoting R&D of the forefront areas:** Build on Japan's space science successes and world-class expertise, such as Moon and asteroid science missions, human space activity (ISS), and a space solar power program.

5. **Foster strategic industries for the 21st century:** "The Government will place the space industry among strategic industries in the 21st century and enhance competitiveness by promoting space machinery smaller, serialized, communized and standardized"; and

6. **Consider the environment:** "The Government will take measures considering the global and space environment, such as [the] space debris issue." The Plan commits Japan to developing the following four systems for utilization:

   - Land and ocean observing satellites
   - Global environmental change and weather observing satellite system
   - Advanced telecommunications satellite system
   - Satellite system for national security and to four R&D programs:

     Space science

     Human space activity

     Space solar power

     Small demonstration satellite

2.3.7.5 European Space Activities

The European Space Research Organization (ESRO) was founded in 1964. France, Britain, and Germany were its leading contributors with seven other initial members: Italy, Belgium, the Netherlands, Sweden, Denmark, Spain, and Switzerland. ESRO's main purpose was to coordinate European space policy and direct research efforts for the peaceful uses of space. Over the following 10 years, the organization developed seven scientific satellites—four that studied the Earth's upper atmosphere and aurorae (the northern and southern lights), two that studied the Earth's magnetic field and the solar wind, and an orbiting ultraviolet observatory. In order to be able to launch their own satellites, the European Launcher Development Organization (ELDO) was set up to coordinate its development. In May 1975, the European powers united all their joint space efforts in a single new organization—the ESA. One of the first satellites launched under the new ESA organization was the gamma-ray astronomy observatory COS-B. The first detailed views of the universe in gamma-rays were provided by the ESA's mission COS-B along with NASA's SAS-2. Launched on August 9, 1975, COS-B was originally projected to last 2 years, but it operated successfully for 6 years and 8 months and provided the first complete map of the galaxy in gamma rays. Since its formation in 1975, ESA has become a major player in the commercial launch industry. It has also developed many ground-breaking satellites and probes and sent astronauts into space in collaboration with another space launcher called Vega for smaller payloads. The organization has also grown in membership from initially 10–17 now. ESA developed a successful series of remote-sensing satellites. Earth Resource Satellite, ERS-1, was among the first to carry powers. ESA developed Ariane 1–5 and is presently developing a solid rocket.

The following provides a very brief information about the European countries participating in ESA:

France

France's national space agency is the Centre National d'Études Spatiales (CNES), which provides partial funding for Arianespace, the commercial entity that manufactures and operates the Ariane launcher. The country's focus has been on access to space, civil applications of space, sustainable development, science and technology research, and security and defense. In 2009, France announced a multi-year strategy to increase funding and utilization of dual-use military and civil space technologies and applications in coordination with other European nations.

Germany

The German Aerospace Center (DLR), which carries out research and development work in aeronautics, space, transportation and energy is Germany's national space agency. The agency focuses on observing the Earth and universe, research for protecting the environment and development of environmentally friendly technologies to promote mobility, communication, and security. Major German space programs have continued to emphasize Earth observation.

Italy

Italy's national space agency is the Agenzia Spaziale Italiana (ASI), which promotes, coordinates, and conducts space activities in Italy, including developing space technology. ASI plays a key role at the European level, where Italy is the third largest country to contribute to ESA. In 2010, Italy launched the fourth, and final, satellite in the COSMO-SkyMed Earth-observation satellite constellation. COSMO-SkyMed provides dual-use radar Earth observation data intended to meet civilian and military needs. ASI is leading Europe's development of the Vega small-class launcher, which is intended to complement the larger Ariane V launcher.

United Kingdom (U.K.)

The U.K. Space Agency (UKSA) was launched in 2010 to replace the British National Space Centre and bring all U.K. civil space activities under one single management. It coordinates U.K. civil space activity, supports academic research, nurtures the U.K. space industry, raises domestic and international awareness of U.K. space activities, works to increase understanding of space science and its practical benefits and inspires the next generation of U.K. scientists and engineers. The U.K. Space Leadership Council provides the UKSA with strategic guidance as it coordinates and promotes U.K. civil space activity in the academic, industrial, scientific, and educational arenas.

2.3.7.6 *Indian Space Activities*

In 1962, the Indian National Committee for Space Research (INCOSPAR) was established which was followed by the ISRO (Indian Space Research Organization) in 1969. ISRO institutionalized space activities in India.[56]

For almost five decades, the Indian space program has been driven by the vision of Dr. Vikram Sarabhai, considered the father of the Indian space program, who stated in part: "If we are to play a meaningful role nationally, and in the community of nations, we must be second to none in the application of advanced technologies to the real problems of man and society". In this and other respects, India has been particularly successful. As India looks toward celebrating 50 years as a space faring nation, its program has expanded beyond its historic focus on space applications toward ever more sophisticated science missions and even the possibility of entering the human spaceflight arena.

India does not have a consolidated national space policy and, according to the Government of India website, its most recent science and technology policy was issued in 2003. Indian academicians and other experts have called for the formulation of a national space policy. Today, India's primarily civil space program is administered by the Department of Space (DoS), which was established in June 1972. The ISRO, the Indian national space agency, is the primary research and development arm of the DoS, and other agencies of the DoS include the Physical Research Laboratory (PRL), the National Atmospheric Research Laboratory (NARL), the North Eastern Space Applications Centre (NE-SAC) and the Semi-Conductor Laboratory (SCL). The Indian Space program under the DoS aims to promote the development and application of space science and technology for the socioeconomic benefit of the country. Between 2003 and 2010, India's space budget increased from 473 million dollars to 1.25 billion dollars annually. Currently, India is ranked sixth among space faring nations after the United States, Europe, Russia, China, and Japan.

India leads the world in "bringing space to the people." It has established programs that bring Earth observation data to the rural villages of India to aid in crop production. ISRO has also set up an enviable network for tele-education and telemedicine services throughout India. ISRO has successfully created capability for national communication, television broadcasting and meteorological services, and an Indian Remote-Sensing Satellites (IRS) system for resources monitoring and management. The Antrix Corporation, the commercial arm of ISRO, sells remote-sensing imagery as well as satellites and launch services.

From the beginning, India was keen on developing its own launch capability so that its space program could operate autonomously. The launch program began with sounding rockets, and the second launch of India's Space Launch Vehicle (SLV-3), a four-stage vehicle, in 1980 led to the development of a five-stage Augmented Satellite Launch Vehicle capable of placing a 150 kg satellite into LEO. This was later followed by the highly successful Polar Satellite Launch Vehicle (PSLV).

Although the PSLV is designed to place IRS into sun-synchronous orbit, it has also placed payloads into geostationary orbit. To date, the PSLV has placed 25 Indian and 27 foreign satellites into orbit on a commercial basis. However, the Geo Synchronous Satellite Launch Vehicle (GSLV-2) has been less successful, and India is now placing its hope on a GSLV-3, which was successfully launched by ISRO on June 5, 2017 from the Satish Dhawan Space Center in Andhra Pradesh. Success with the GSLV-3 will enable India to launch its own communications satellites that are currently launched commercially by an Ariane-5. The GSLV-3 will also enable India to pursue its lunar science mission program. With the help of Indian government backing, ISRO announced in August 2007 that it intended to develop a human spaceflight program. However, after a change of leadership at ISRO, the program has languished or in the words of its previous chief, "the new management is not doing enough to push it with the government and get its approval". The program was started with the development and launch of the Space Capsule Recovery Experiment in January 2007.

Due to a strong interest within the Indian scientific community, ISRO has begun implementing dedicated science missions. In October 2008, India joined the ranks of China and Japan by launching Chandrayaan-1, the first in a series of lunar orbiters. Chandrayaan-1 is a complex spacecraft with 11 instruments including instruments from Europe (ESA), United States (NASA), and Bulgaria. The mission, designed to measure the chemical and mineralogical properties of the Moon, particularly the southern pole, enjoyed a huge success, and was instrumental in the discovery of water on the Moon. ISRO has plans for a dedicated astronomy mission (Astrosat), a solar mission (Aditya-1), and a second Chandrayaan mission to the Moon was originally planned for 2013 but is now planned for first quarter of 2018. However, the timing for Chandrayaan-2 was somewhat dependent on the successful development of GSLV-3, which was successfully launched only recently as mentioned above or a decision to procure the launch. In the future, ISRO plans to undertake a dedicated space science mission every 3–5 years.

During a Space Science and Security Conference held in New Delhi in January 2011, it was highlighted that although India had only a civilian space program, the inherent dual-use nature of satellite technology could be adapted for military purposes as well. Indeed, the dependence of the Indian armed forces on space technology is an undisputed fact. However, no specific information in this regard is available to quantify it precisely.

Indian analysts are of the opinion that presently, this use is limited to areas like communication and navigation, but in the future, the dependence is likely to increase for surveillance and even for counter-space capabilities. For now, India has already established a "Space Cell" under the command of the Integrated Defense Services Headquarters. This is a single tri-service window for interaction in space by all agencies, including external ones. It also acts as a single organization for integration among the armed forces, the DoS, and ISRO.

India's government space program is structured around the ISRO, which manages civil space programs. India's space program has generally focused on technological, scientific, and social development through space capabilities. Only recently has the country migrated toward programs such as space exploration and military applications. India often emphasizes its strategy of international collaboration with leading space powers and has strengthened its relationships with Europe in its efforts to commercialize its launch capability. India has also worked with Russia to develop its own geostationary launch vehicle, and has successfully launched its own spacecraft to the Moon in 2008.

The space policies for the three major contributors to the ESA, namely Germany, France, and Italy are summarized below. The UK has also been included since the UK space sector is gaining political visibility and support. For the most part, the space programs of ESA Member States are increasingly being influenced by the overall European space policies.

The 2016/2017 Performance Indicators are shown in the following Figure 2.44:

2.3.7.7 Other Countries' Space Activities

2.3.7.7.1 Canada

Government programs include the Canadian Space Agency (CSA), the Canadian Center for Remote Sensing and the Canadian Forces. Canada employs a niche strategy, focusing on expertise in three areas: space robotics, radar technology for Earth observation, and advanced satellite communications. Budgets focus heavily on space science and Earth observations, as well as human spaceflight. Canada's space program is uniquely tied to the ESA and U.S. civilian and military programs. Canada has a cooperation agreement with ESA, collaborates on many NASA missions and is a contributor to the International Space Station (ISS). Canada was the first country to have its own satellite in orbit. Alouette 1 was launched by a U.S. Thor-Agena B rocket in September 1962, which was followed by a series of similar satellites as well as national projects such as the ANIK satellite which was the first domestic satellite launched in 1972. RADARSAT was also the first active remote sensing satellite that was developed and launched by Canada in 1995 using SAR (Synthetic Aperture Radar). More on RADARSAT is covered in the next chapter.

The Key Performance Indicators are shown in bold

Outcome 1 : We will have clear and established space policies and policy positions.

No.	PI	Metric
1.1	Using and continuing to develop our evidence on the impact that the Agency has on the UK space sector, working with stakeholders to maximise the value we generate for the UK	Publish the new Size & Health report by end of Q3, incorporating the improved methodology Complete evaluation reports for 3 programmes by end of Q4
1.2	Establish Ministerial Committee to address strategic space security and prosperity discussions across Whitehall	Agree top three issues for Ministers across Government to prioritise by end of Q3 Co-ordinate agreed ToR across Whitehall & implement supporting governance structures by Q3
1.3	Set out the priorities for the UK space sector addressing the need for continued economic growth, increased exports and industrial sustainability	Civil space strategy 2016-2020 in place by Q3 setting out government plans for the sector within the 5 year timeframe
1.4	Provide national and international leadership in EO policy and strategy	Complete delivery of the implementation plan of the Earth observation strategy by the end of Q4
1.5	Coordinate and develop international engagements likely to secure UK priorities in UK exports and inwards investment	Demonstrate effective links with 50% of emerging space economies by end of Q4 Develop international guidelines as part of all future International Partnership Programme calls In partnership with UKTI and Industry, promote exports into 4 key markets by end of Q4

Outcome 2 : UK space policies and policy positions will be effectively represented at national and international level.

No.	PI	Metric
2.1	Secure UK goals in the European Commission's space strategy through continuing to foster positive relationships with the Commission	Develop UK objectives for the European Space Policy by end of Q1 to coincide with the start of the Commission's expected consultation process Influence the European Space Policy to best reflect UK objectives by end of Q2
2.2	Ensure that the UK is prepared for the space aspects of the UK's presidency of the European Council in 2017	Agree a plan of activities to promote European space policy by end of Q3
2.3	Enable the UK to further develop climate services from space data	Establish the UK Space Agency roles and funding requirements by end of Q1
2.4	Maintain an internationally competitive UK space regulatory regime	Formalise proposals for an outline regulatory framework for Earth Observation data management by end of Q4 Implement provisions for third party liability insurance requirements for satellite constellation operators by end of Q4
2.5	Commence the reform of the economic cost of delivering the space licensing regime and the fees charged to ensure value for money for the tax-payer and a clear fee system for applicants	Issue consultation on possible fees reform by end of Q2 Publish government response to consultation by end of Q4
2.6	Identify an early adopter of the Public Regulated Service (PRS) component of Galileo	Demonstrate the initial concept for regulatory oversight of PRS regulation by end of Q4

Outcome 3 : The UK will maintain and grow its national capability in space.

No.	PI	Metric
3.1	Set out and achieve the UK programme priorities for investment at the European Space Agency's Council of Ministers in December 2016	Approval of business cases and allocation of financial resource to support UK CMin 16 objectives by end of November 2016
3.2	Establish the initial capability to deliver early EU Space Surveillance and Tracking (SST) services by 1 June 2016	Establish initial Space Surveillance services to Galileo and Skynet under the EU SST programme by end of Q2 Establish a coordination mechanism with the MOD for civil / military SST by end of Q4
3.3	Identify those assets within the space sector which, if impaired, would most critically impact national resilience	Establish a database of UK space sector assets and assess the criticality of their contribution to national infrastructure by end of Q4
3.4	Further develop the UK's approach to managing space related spectrum with partners across Government, industry and in key international forums	Establish an initial UK sector position on the World Radio Conference 2019 by end of Q4 Through the Central Management Unit, reassess spectrum charges and consider potential spectrum releases against HMT targets by end of Q4
3.5	Deliver a programme of growth activities that supports the development of a network of local space clusters linked with each other and the UK Space Gateway at Harwell	By Q2, support the development of a scheme that better enables companies to access the facilities and expertise available at the UK Space Gateway at Harwell. By end of Q4, eight companies to have used the scheme By end of Q2, conduct an analysis of the new Size & Health data to establish baselines and trends of space activity across the UK. Use this information to support engagement with LEPs and Devolved Administrations By end of Q4, actively support the development of three space regional Centres of Excellence and six space incubation projects that engage with and support companies in both upstream and downstream space sector
3.6	Run a National Spaceflight Programme to deliver a stepwise approach to establishing a commercial small satellite launch capability in the UK as set out in the National Space Policy. Initial capability will build on the operation of sub-orbital science spaceflights from a UK spaceport	Set out the actions to address UK priorities for National Spaceflight by the end of Q2
3.7	Engage with the space industrial sector to facilitate the growth of existing and new SMEs in the UK	Work with industry to ensure that the UK SME community are engaged with UK Space Agency activities by end Q4
3.8	Make the UK a key global access point for the exploitation of sentinel and other EO data	Develop by the end of Q4 a plan to sustain the provision of data through the EO ground segment

FIGURE 2.44
Performance indicators.

(*Continued*)

Outcome 4 : UK investment in space will be effective, targeted and delivers tangible economic, societal or social benefit.

No.	PI	Metric
4.1	Fund and monitor the progress of the development and delivery of the Agency's agreed national and international space programmes	Ensure all projects remain within approved performance, time and cost parameters
4.2	Increase in UK research facilities made available to the European Programme for Life and Physical Science (ELIPS) infrastructure	Ensure UK research facilities are included by ESA in the ELIPS programme by end of Q3
4.3	Maintain progress against performance, time and cost against the approved parameters for the NovaSAR project	Spacecraft flight readiness review carried out by end of Q2
4.4	Maintain progress against performance, time and cost against the approved parameters for the Synergetic Air-Breathing Rocket Engine (SABRE) project	Successful completion of the Sub-system Baseline Design Reviews by end of Q4
4.5	Successfully deliver the International Partnership Programme (IPP) to performance, time and cost	Define and agree a costed 2 year plan by end of Q1
		Place relevant calls to industry needed to deliver the plan by end of Q3
		Ensure in-year financial spending committed in accordance with the plan by end of Q4
4.6	Continue to provide leadership on space technology development across the UK via the National Space Technology Programme (NSTP)	Commitment to run the Fast Track call by end of Q3
4.7	Consider options for a more resilient and effective space weather operational capability to support BIS in mitigating Government's space weather risk	Scope and develop the evidence base to assess the value of a UK led international space weather mission, delivering initial analysis by end of Q4
4.8	Maintain progress against performance, time and cost against the approved parameters for the development of national space propulsion facilities	Successful completion of Facility Acceptance and Final Review for the Northern Ireland facility by end of Q4
		Successful completion of upgrade of medium altitude testing facility at Westcott by end of Q4

Outcome 5 : The criticality and utility of the space sector to science, enterprise and economic growth will be increasingly understood by policy makers, commerce and the general public.

No.	PI	Metric
5.1	Through the SSGP, facilitate the public sector in using satellite enabled services for smarter, more efficient operations, in addition to stimulating economic growth	Space enabled Applications Plan published by end of Q3
		Deliver a minimum of two thematic roadmaps which involve at least 8 Government organisations and Departments, focussing on how Space can save the UK money, and grow outputs by end of Q4
		Progress work on cross Government products including a Space Catalogue and a joined up Space training package delivered by end of Q4
5.2	Exploit the unique opportunity of the Principia Mission, Farnborough and a Science is GREAT campaign to deliver increased awareness of UK space activities, inspirational education programmes, its relevance to everyday life and opportunities for UK industry	Evaluate UK perceptions of space activity against the 2015/16 benchmark
		Increase the total number of engagements by young people with the Principia education programme to a total of 1,000,000 by end of Q4
5.3	Deliver an effective formal and informal education programme for 2016/17	Engage over 140,000 young people in formal and informal space education by end of Q4
		Plan education campaigns to capitalise on upcoming missions such as ExoMars and James Webb Space Telescope by end of Q3
5.4	Support the ambition of the UK space sector that STEM graduates possess the appropriate skills and attributes	A careers website will be operational by end of Q4
		A skills portal will be hosted by the Open University by end of Q4

Outcome 6 : The UK Space Agency will have the capability, capacity and culture to deliver the Civil Space Strategy.

No.	PI	Metric
6.1	Develop a workforce plan to enable the Agency to best respond to BIS2020 challenges and to address on-going staff retention risk	Workforce plan incorporating a succession plan developed by end of Q1
6.2	Continue to develop staff capability and engagement	Ensure each employee has a personal development plan detailing at least 5 L&D days a year
		Implement the L&D strategy by end of Q4
		Implement the Staff Survey Action Plan by the end of Q2
6.3	Deliver operational budget efficiencies to contribute to the BIS 2020 targets	Deliver operational budget efficiencies of £20k
6.4	Agency programme, capital and administration budgets for 2016/17 are managed to deliver corporate plan objectives	Agency agreed outturn is within budget (tolerance of +0 and -1%)
6.5	Produce the 2015/16 Annual Report and Accounts	NAO provide an unqualified honest opinion on the 15/16 Accounts by end of Q1
		2015/16 Annual Report and Accounts laid by 30 June 2016

FIGURE 2.44 (CONTINUED)
Performance indicators.

2.3.7.7.2 Iran

Iranian space activities are coordinated by the Iranian Space Agency (ISA) which was established on February 1, 2004 and focus on developing communication and remote-sensing satellites, launch vehicle technology, and astronomy research. In 2009, ISA launched its first research and telecommunication satellite. In 2010, ISA launched living organisms into space onboard its experimental biological capsule. Iran has developed an expendable satellite launch vehicle named Safir SLV. Measuring 22 m in height with a core diameter of 1.25 m, with two liquid propellant stages, a single thrust chambered first stage and a two-thrust chambered, step-throttled second stage, the SLV has a lift off mass exceeding 26 tons. The Safir-1B is the second generation of Safir SLV and can carry a satellite weighing 60 kg into an elliptical orbit of 300–450 km. The thrust of the Safir-1B rocket engine has been increased from 32 to 37 tons. In 2010, a more powerful rocket named Simorgh (Phoenix) was built. Its mission is to carry heavier satellites into orbit. The Simorgh rocket is 27 m (89 ft) long, and has a mass of 77 tonnes (85 tons). On February 2, 2013, the head of the Iranian space agency, Fazeli mentioned that the new satellite launch vehicle, Qoqnoos will be used after the Simorgh SLV for heavier payloads.

Fajr, is an imaging satellite which also carries an experimental locally made GPS system built by Iran Electronics Industries. The satellite will have a life span of 1.5 years and an imaging resolution of 500–1000 m. It is the first Iranian satellite to use "Cold Gas Thruster" and has solar panels. Originally, it was to be launched in 2012. As were alleged, non-announced by Iran two failed launches of Fajr satellites occurred on May 23 and October 2012. Finally, Fajar was successfully launched and placed into orbit on February 2, 2015.

Now that we have looked at some of the space-related activities of the countries that are participant in space activities more than others, it is interesting to note the key principles that are important for international coordination for sustainable space exploration and examples of resulting requirements for the mechanism.

Principles	Resulting Requirements
Open and Inclusive	* Receives inputs from all interested agency participants that invest in and perform activities related to space exploration. *Provides for consultations among all interested agencies with a vested interest in space exploration, and also space agencies or national government agencies without specific related capabilities
Flexible and Evolutionary	• takes into account and may integrate existing consultation and coordination mechanisms • allows consultation and coordination structures and mechanism(s) to gradually build and evolve as requirements for these activities grow • allows for entry of assigned representatives of governments with a vested interest and clear stake in space exploration • provides for different levels of consultation and coordination
Effective	• encourages participating agencies to accept the role of the coordination process and act upon the anticipated results of the coordination mechanism
Mutual Interest	• contributes to common peaceful goals and benefits all participants • respects the national prerogatives of participating agencies • allows for optional participation based on the level of each agency's interest

FIGURE 2.45
Key principles that are important for international coordination for sustainable space Exploration.[57]

UN Documents
Gathering a body of global agreements

Home | *Sustainable Development* | *Education* | *Water* | *Culture of Peace* | *Human Rights* | *Keywords* | *Search*

<table>
<tr><td>United Nations</td><td>**A**/RES/18/1962</td></tr>
</table>

 General Assembly

Distr: General
13 December 1963
Original: English

Eighteenth session
Agenda item 28 a

RESOLUTION ADOPTED BY THE GENERAL ASSEMBLY

1962 (XVIII). Declaration of Legal Principles Governing the Activities of States in the Exploration and Use of Outer Space

The General Assembly,

Inspired by the great prospects opening up before mankind as a result of man's entry into outer space,

Recognizing the common interest of all mankind in the progress of the exploration and use of outer space for peaceful purposes,

Believing that the exploration and use of outer space should be carried on for the betterment of mankind and for the benefit of States irrespective of their degree of economic or scientific development,

Desiring to contribute to broad international cooperation in the scientific as well as in the legal aspects of exploration and use of outer space for peaceful purposes,

Believing that such co-operation will contribute to the development of mutual understanding and to the strengthening of friendly relations between nations and peoples,

Recalling its *resolution 110 (II)* of 3 November 1947, which condemned propaganda designed or likely to provoke or encourage any threat to the peace, breach of the peace, or act of aggression, and considering that the aforementioned resolution is applicable to outer space,

Taking into consideration its resolutions 1721 (XVI) of 20 December 1961 and 1802 (XVII) of 14 December 1962, adopted unanimously by the States Members of the United Nations,

Solemnly declares that in the exploration and use of outer space States should be guided by the following principles:

1. The exploration and use of outer space shall be carried on for the benefit and in the interests of all mankind.

2. Outer space and celestial bodies are free for exploration and use by all States on a basis of equality and in accordance with international law.

3. Outer space and celestial bodies are not subject to national appropriation by claim of sovereignty, by means of use or occupation, or by any other means.

4. The activities of States in the exploration and use of outer space shall be carried on in accordance with international law, including the Charter of the United Nations, in the interest of maintaining international peace and security and promoting international cooperation and understanding.

5. States bear international responsibility for national activities in outer space, whether carried on by governmental agencies or by non-governmental entities, and for assuring that national activities are carried on in conformity with the principles set forth in the present Declaration. The activities of non-governmental entities in outer space shall require authorization and continuing supervision by the State concerned. When activities are carried on in outer space by an international organization, responsibility for compliance with the principles set forth in this Declaration shall be borne by the international organization and by the States participating in it.

6. In the exploration and use of outer space, States shall be guided by the principle of co-operation and mutual assistance and shall conduct all their activities in outer space with due regard for the corresponding interests of other States. If a State has reason to believe that an outer space activity or experiment planned by it or its nationals would cause potentially harmful interference with activities of other States in the peaceful exploration and use of outer space, it shall undertake appropriate international consultations before proceeding with any such activity or experiment. A State which has reason to believe that an outer space activity or experiment planned by another State would cause potentially harmful interference with activities in the peaceful exploration and use of outer space may request consultation concerning the activity or experiment.

7. The State on whose registry an object launched into outer space is carried shall retain jurisdiction and control over such object, and any personnel thereon, while in outer space. Ownership of objects launched into outer space, and of their component parts, is not affected by their passage through outer space or by their return to the earth. Such objects or component parts found

beyond the limits of the State of registry shall be returned to that State, which shall furnish identifying data upon request prior to return.

8. Each State which launches or procures the launching of an object into outer space, and each State from whose territory or facility an object is launched, is internationally liable for damage to a foreign State or to its natural or juridical persons by such object or its component parts on the earth, in air space, or in outer space.

9. States shall regard astronauts as envoys of mankind in outer space, and shall render to them all possible assistance in the event of accident, distress, or emergency landing on the territory of a foreign State or on the high seas. Astronauts who make such a landing shall be safely and promptly returned to the State of registry of their space vehicle.

1280th plenary meeting
13 December 1963

UN Documents
Gathering a body of global agreements

Home | Sustainable Development | Education | Water | Culture of Peace | Human Rights | Keywords | Search

United Nations **A**/RES/21/2222

General Assembly

Distr: General
19 December 1966
Original: English

Twenty-first session
Agenda item 30

The General Assembly,

Having considered the report of the Committee on the Peaceful Uses of Outer Space covering its work during 1966,/10 and in particular the work accomplished by the Legal Sub-Committee during its fifth session, held at Geneva from 12 July to 4 August and at New York from 12 September to 16 September,

Noting further the progress achieved through subsequent consultations among States Members of the United Nations,

Reaffirming the importance of international co-operation in the field of activities in the peaceful exploration and use of outer space, including the Moon and other celestial bodies, and the importance of developing the rule of law in this new area of human endeavour,

1. *Commends* the Treaty on Principles Governing the Activities of States in the Exploration and Use of Outer Space, including the Moon and Other Celestial Bodies, the text of which is annexed to the present resolution;

2. *Requests* the Depositary Governments to open the Treaty for signature and ratification at the earliest possible date;

3. *Expresses* its hope for the widest possible adherence to this Treaty;

4. *Requests* the Committee on the Peaceful Uses of Outer Space:

   a. To continue its work on the elaboration of an agreement on liability for damages caused by the launching of objects into outer space and an agreement on assistance to and return of astronauts and space vehicles, which are on the agenda of the Committee;

   b. To begin at the same time the study of questions relative to the definition of outer space and the utilization of outer space and celestial bodies, including the various implications of space communications;

   c. To report on the progress of its work to the General Assembly at its twenty-second session.

1499 plenary meeting
19 December 1966

ANNEX

Treaty on Principles Governing the Activities of States in the Exploration and Use of Outer Space, including the Moon and Other Celestial Bodies

The States Parties to this Treaty,

Inspired by the great prospects opening up before mankind as a result of man's entry into outer space,

Recognizing the common interest of all mankind in the progress of the exploration and use of outer space for peaceful purposes,

Believing that the exploration and use of outer space should be carried on for the benefit of all peoples irrespective of the degree of their economic or scientific development,

Desiring to contribute to broad international cooperation in the scientific as well as the legal aspects of the exploration and use of outer space for peaceful purposes,

Believing that such cooperation will contribute to the development of mutual understanding and to the strengthening of friendly relations between States and peoples,

Recalling resolution 1962 (XVIII), entitled "Declaration of Legal Principles Governing the Activities of States in the Exploration and Use of Outer Space", which was adopted unanimously by the United Nations General Assembly on 13 December 1963,

Recalling resolution 1884 (XVIII), calling upon States to refrain from placing in orbit around the Earth any objects carrying nuclear weapons or any other kinds of weapons of mass destruction or from installing such weapons on celestial bodies, which was adopted unanimously by the United Nations General Assembly on 17 October 1963,

Taking account of United Nations General Assembly *resolution 110 (II)* of 3 November 1947, which condemned propaganda designed or likely to provoke or encourage any threat to the peace, breach of the peace or act of aggression, and considering that the aforementioned resolution is applicable to outer space,

Convinced that a Treaty on Principles Governing the Activities of States in the Exploration and Use of Outer Space, including the Moon and Other Celestial Bodies, will further the purposes and principles of the Charter of the United Nations,

Have agreed on the following:

Article I

The exploration and use of outer space, including the Moon and other celestial bodies, shall be carried out for the benefit and in the interests of all countries, irrespective of their degree of economic or scientific development, and shall be the province of all mankind.

Outer space, including the Moon and other celestial bodies, shall be free for exploration and use by all States without discrimination of any kind, on a basis of equality and in accordance with international law, and there shall be free access to all areas of celestial bodies.

There shall be freedom of scientific investigation in outer space, including the Moon and other celestial bodies, and States shall facilitate and encourage international cooperation in such investigation.

Article II

Outer space, including the Moon and other celestial bodies, is not subject to national appropriation by claim of sovereignty, by means of use or occupation, or by any other means.

Article III

States Parties to the Treaty shall carry on activities in the exploration and use of outer space, including the Moon and other celestial bodies, in accordance with international law, including the *Charter of the United Nations*, in the interest of maintaining international peace and security and promoting international cooperation and understanding.

Article IV

States Parties to the Treaty undertake not to place in orbit around the Earth any objects carrying nuclear weapons or any other kinds of weapons of mass destruction, install such weapons on celestial bodies, or station such weapons in outer space in any other manner.

The Moon and other celestial bodies shall be used by all States Parties to the Treaty exclusively for peaceful purposes. The establishment of military bases, installations and fortifications, the testing of any type of weapons and the conduct of military manoeuvres on celestial bodies shall be forbidden. The use of military personnel for scientific research or for any other peaceful purposes shall not be prohibited. The use of any equipment or facility necessary for peaceful exploration of the Moon and other celestial bodies shall also not be prohibited.

Article V

States Parties to the Treaty shall regard astronauts as envoys of mankind in outer space and shall render to them all possible assistance in the event of accident, distress, or emergency landing on the territory of another State Party or on the high seas. When astronauts make such a landing, they shall be safely and promptly returned to the State of registry of their space vehicle.

In carrying on activities in outer space and on celestial bodies, the astronauts of one State Party shall render all possible assistance to the astronauts of other States Parties.

States Parties to the Treaty shall immediately inform the other States Parties to the Treaty or the Secretary-General of the United Nations of any phenomena they discover in outer space, including the Moon and other celestial bodies, which could constitute a danger to the life or health of astronauts.

Article VI

States Parties to the Treaty shall bear international responsibility for national activities in outer space, including the Moon and other celestial bodies, whether such activities are carried on by governmental agencies or by non-governmental entities, and for assuring that national activities are carried out in conformity with the provisions set forth in the present Treaty. The activities of non-governmental entities in outer space, including the Moon and other celestial bodies, shall require authorization and continuing supervision by the appropriate State Party to the Treaty. When activities are carried on in outer space, including the Moon and other celestial bodies, by an international organization, responsibility for compliance with this Treaty shall be borne both by the international organization and by the States Parties to the Treaty participating in such organization.

Article VII

Each State Party to the Treaty that launches or procures the launching of an object into outer space, including the Moon and other celestial bodies, and each State Party from whose territory or facility an object is launched, is internationally liable for damage to another State Party to the Treaty or to its natural or juridical persons by such object or its component parts on the Earth, in air space or in outer space, including the Moon and other celestial bodies.

Article VIII

A State Party to the Treaty on whose registry an object launched into outer space is carried shall retain jurisdiction and control over such object, and over any personnel thereof, while in outer space or on a celestial body. Ownership of objects launched into outer space, including objects landed or constructed on a celestial body, and of their component parts, is not affected by their presence in outer space or on a celestial body or by their return to the Earth. Such objects or component parts found beyond the limits of the State Party to the Treaty on whose registry they are carried shall be returned to that State Party, which shall, upon request, furnish identifying data prior to their return.

Article IX

In the exploration and use of outer space, including the Moon and other celestial bodies, States Parties to the Treaty shall be guided by the principle of cooperation and mutual assistance and shall conduct all their activities in outer space, including the Moon and other celestial bodies, with due regard to the corresponding interests of all other States Parties to the Treaty.

States Parties to the Treaty shall pursue studies of outer space, including the Moon and other celestial bodies, and conduct exploration of them so as to avoid their harmful contamination and also adverse changes in the environment of the Earth resulting from the introduction of extraterrestrial matter and, where necessary, shall adopt appropriate measures for this purpose. If a State Party to the Treaty has reason to believe that an activity or experiment planned by it or its nationals in outer space, including the Moon and other celestial bodies, would cause potentially harmful interference with activities of other States Parties in the peaceful exploration and use of outer space, including the Moon and other celestial bodies, it shall undertake appropriate international consultations before proceeding with any such activity or experiment.

A State Party to the Treaty which has reason to believe that an activity or experiment planned by another State Party in outer space, including the Moon and other celestial bodies, would cause potentially harmful interference with activities in the peaceful exploration and use of outer space, including the Moon and other celestial bodies, may request consultation concerning the activity or experiment.

Article X

In order to promote international cooperation in the exploration and use of outer space, including the Moon and other celestial bodies, in conformity with the purposes of this Treaty, the States Parties to the Treaty shall consider on a basis of equality any requests by other States Parties to the Treaty to be afforded an opportunity to observe the flight of space objects launched by those States.

The nature of such an opportunity for observation and the conditions under which it could be afforded shall be determined by agreement between the States concerned.

Article XI

In order to promote international cooperation in the peaceful exploration and use of outer space, States Parties to the Treaty conducting activities in outer space, including the Moon and other celestial bodies, agree to inform the Secretary-General of the United Nations as well as the public and the international scientific community, to the greatest extent feasible and practicable, of the nature, conduct, locations and results of such activities. On receiving the said information, the Secretary-General of the United Nations should be prepared to disseminate it immediately and effectively.

Article XII

All stations, installations, equipment and space vehicles on the Moon and other celestial bodies shall be open to representatives of other States Parties to the Treaty on a basis of reciprocity. Such representatives shall give reasonable advance notice of a projected visit, in order that appropriate consultations may be held and that maximum precautions may be taken to assure safety and to avoid interference with normal operations in the facility to be visited.

Article XIII

The provisions of this Treaty shall apply to the activities of States Parties to the Treaty in the exploration and use of outer space, including the Moon and other celestial bodies, whether such activities are carried on by a single State Party to the Treaty or jointly with other States, including cases where they are carried on within the framework of international intergovernmental organizations.

Any practical questions arising in connection with activities carried on by international intergovernmental organizations in the exploration and use of outer space, including the Moon and other celestial bodies, shall be resolved by the States Parties to the Treaty either with the appropriate international organization or with one or more States members of that international organization, which are Parties to this Treaty.

Article XIV

1. This Treaty shall be open to all States for signature. Any State which does not sign this Treaty before its entry into force in accordance with paragraph 3 of this article may accede to it at any time.

2. This Treaty shall be subject to ratification by signatory States. Instruments of ratification and instruments of accession shall be deposited with the Governments of the Union of Soviet Socialist Republics, the United Kingdom of Great Britain and Northern Ireland and the United States of America, which are hereby designated the Depositary Governments.

3. This Treaty shall enter into force upon the deposit of instruments of ratification by five Governments including the Governments designated as Depositary Governments under this Treaty.

4. For States whose instruments of ratification or accession are deposited subsequent to the entry into force of this Treaty, it

shall enter into force on the date of the deposit of their instruments of ratification or accession.

5. The Depositary Governments shall promptly inform all signatory and acceding States of the date of each signature, the date of deposit of each instrument of ratification of and accession to this Treaty, the date of its entry into force and other notices.

6. This Treaty shall be registered by the Depositary Governments pursuant to *Article 102 of the Charter of the United Nations.*

Article XV

Any State Party to the Treaty may propose amendments to this Treaty. Amendments shall enter into force for each State Party to the Treaty accepting the amendments upon their acceptance by a majority of the States Parties to the Treaty and thereafter for each remaining State Party to the Treaty on the date of acceptance by it.

Article XVI

Any State Party to the Treaty may give notice of its withdrawal from the Treaty one year after its entry into force by written notification to the Depositary Governments. Such withdrawal shall take effect one year from the date of receipt of this notification.

Article XVII

This Treaty, of which the Chinese, English, French, Russian and Spanish texts are equally authentic, shall be deposited in the archives of the Depositary Governments. Duly certified copies of this Treaty shall be transmitted by the Depositary Governments to the Governments of the signatory and acceding States.

In witness whereof the undersigned, duly authorized, have signed this Treaty.

Done in triplicate, at the cities of London, Moscow and Washington, D.C., the ... day of ..., one thousand nine hundred and ... ./11

Note
10/ See Official Records of the General Assembly, Twenty-first session, Annexes, agenda items 30, 89 and 91, document A/6431.
11/ The Treaty was signed in London, Moscow and Washington on 27 January, 1967.

R. H. Goddard
Rocket Apparatus
Application filed October 1, 1913.

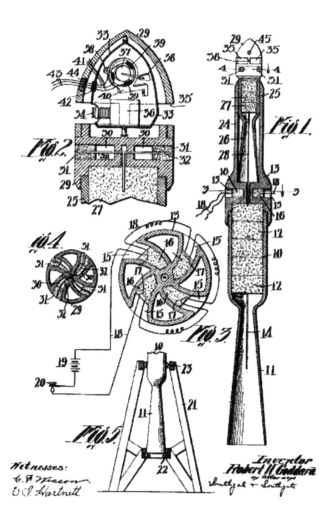

2.4 Questions

1. Describe Newton's first, second, and third laws of motion.
2. Describe the Copernicus theorem on the revolution of the heavenly spheres.
3. To what altitude does the Earth's atmosphere reaches according to NASA's definition?
4. Describe the layers of the Earth's atmosphere and their unique characteristics.
5. What areas does the space law encompass and when and why such laws began to be enforced globally through ITU conventions?

6. Explain the purposes of the UN's COPUOS.

7. Explain the goal of the ITU/WARC of 1985 and 1988 and the conclusion they arrived at.

8. Explain the key findings of the May 2007 Global Exploration Strategy.

9. Who was Arthur C. Clarke and what was his main contribution to the space science and technology?

10. Explain the outer space challenges ahead.

11. Explain the essence of the International Agreement on a Code of Conduct for Outer Space Activities.

12. Name the nine indications of the Space Security Index.

13. Describe the nine space vulnerabilities that were adopted in fall 2004.

14. Explain the two categories of the launching vehicles.

15. Briefly describe the Nanosatellites and Microsatellites and their applications.

16. Describe the different types of rocket engines used in satellite communications.

17. How many orbits does it take before the ISS passes over the same point on the equator?

18. Given the way we the humans have behaved on our own planet, regarding the use or misuse of Earth's natural resources, preserving the environment, or lack of, for future generations, a logical question is whether or not we should be encouraging private space exploration. Write an essay expressing your opinion answering this question.

References

1. Othman, E. 2006. Binomial coefficients and Nasir-al-Tusi. *Scientific Research and Essay*, 1(2): 28–32.
2. Al-Todhkira fiilm al-hay'a, See also Boyer (1947) and Dreyer (1953).
3. National Aeronautics and Space Administration (NASA). www.nasa.gov.
4. Zubrin, R. 1999. *Entering Space: Creating a Spacefaring Civilization*. New York: Penguin Group, Supra note 54, at 12, www.amazon.com/Entering-Space-Creating-Spacefaring-Civilization/dp/1585420360#reader_1585420360.
5. Resolution 1348 (XII): Question of the peaceful uses of outer space. Adopted by the General Assembly of the United Nation's 792nd plenary meeting, December 13, 1958.
6. Resolution 1472 (XIV): International co-operation in the peaceful uses of outer space. Adopted Resolution 1472 by the General Assembly of the United Nations, December 12, 1959.

7. Resolution 1721 (XVI): International co-operation in the peaceful uses of outer space. Adopted by the General Assembly of the United Nations, December 20, 1961.
8. Resolution 3235 (XXIX): Convention on the registration of objects launched into outer space. Adopted by the General Assembly of the United Nations, November 12, 1974.
9. Addendum 3, Revision 2 of General Assembly, January 1, 2010.
10. Resolution 1962 (XVIII): Declaration of legal principles governing all activities of states in the exploration and use of outer space. Adopted by the General Assembly of the United Nations, December 1963.
11. Resolution 2222 (XXI): Treaty on principles governing the activities of states in exploration and use of outer space, including the Moon and other celestial bodies, 1966.
12. Kopal, V. 2008. Treaty on principles governing the activities of states in the exploration and use of outer space, including the Moon and other celestial bodies, United Nations, http://untreaty.un.org/cod/avl/pdf/ha/tos/tos_e.pdf.
13. Reynolds, G.H., and R.P. Merges. 1989. *Outer Space: Problems of Law and Policy*. Boulder, CO: Westview Press.
14. Cooper, S.B. 1990. Outer space law. *Harvard Journal of Law and Technology*, Spring: 275–278.
15. United Nations/Thailand/European Space Agency (ESA) Workshop on Space Law. 2010. Activities of states in outer space in light of new developments: Meeting international responsibilities and establishing national legal and policy frameworks. Bangkok, Thailand, November 16–19.
16. https://sustainabledevelopment.un.org/intergovernmental/csd16.
17. www.lpi.usra.edu/lunar/strategies/ISECGLunarRefArchitectureJuly2010.pdf.
18. www.nasa.gov/content/just-released-updates-to-the-global-exploration-roadmap.
19. Highfield, R. 2001. Colonies in space may be only hope, says Hawking, Science Editor, *The Telegraph*, October 16.
20. Clarke, A.C. 1950. *Interplanetary Flight—An Introduction to Astronautics*. New York: Harper and Brothers, Chapter 10.
21. Wikipedia. Space exploration, http://en.wikipedia.org/wiki/Space_exploration.
22. Wikipedia. Timeline of solar system exploration, http://en.wikipedia.org/wiki/Timeline_of_Solar_System_exploration.
23. 2222 (XXI). Treaty on principles governing the activities of states in the exploration and use of outer space, including the Moon and other celestial bodies, www.oosa.unvienna.org/oosa/SpaceLaw/gares/html/gares_21_2222.html.
24. Report of the Scientific and Technical Subcommittee on its 47th Session, Vienna, February 8–19, 2010.
25. Report of the Third UN Conference on the Exploration and Peaceful Use of Outer Space, Vienna, July 19 to 30, 1999.
26. Goddard, R.H. 1914. Patent 1,102,653, Rocket apparatus, July 7, www.google.com/patents?id=ikBJAAAAEBAJ&printsec=abstract&zoom=4&source=gbs_overview_r&cad=0#v=onepage&q&f=false.
27. Secure World Foundation. 2008. Earth and space security: Progress and challenges ahead, December 31, www.newswise.com/articles/earth-and-space-security-progress-and-challenges-ahead.

28. Doyle, R.J. 2003. *Challenges and Opportunities for Information Technology on Future Space Missions*. NASA/JPL, Presentation, RJD 10/30/02, April 1.
29. Scheiderer, G. 2010. Five challenges for future space exploration. *Seattle Astronomy Examiner*, April 20.
30. Preston, R. and J. Baker. 2002. Space challenges. In *United States Air and Space Power in the 21st Century Strategic Appraisal* (Chapter 5, pp. 143–185). Edited by Z. Khalilzad and J. Shapiro. Santa Monica, CA: RAND.
31. Secure World Foundation. 2008. A code of conduct for outer space: A step forward on managing satellite traffic, March 28.
32. Long-term sustainability of space activities, preliminary reflections, UNCOPUOS Scientific and Technical Subcommittee, February 2010.
33. Presentation by H. Klinkrad of the European Space Agency to the Scientific and Technical Subcommittee of the UNOPUOS, February 2008.
34. Weeden, B. and B.B. Walker. 2007. Space traffic management, p. 13, http://swfound.org/media/5167/spacetrafficmgmt-bw-2007.pdf.
35. Space Security Index. 2010, www.spacesecurity.org/spacesecurity.2010.reduced.pdf.
36. Marshall, W. 2008. Space security ways forward, SETI Institute, California, October 12.
37. Tabulation of "Orbital Box Score Data," *Orbital Debris Quarterly* (January 2010) www.orbitaldebris.JSC.nasa.gov/newsletter/pdfs/ODQNv14i1.pdf.
38. www.ulalaunch.com/products_atlasv.aspx.
39. www.faa.gov/about/office_org/headquarters_offices/ast/media/2016_Compendium.pdf.
40. www.faa.gov/about/office_org/headquarters_offices/ast/media/2016_Compendium.pdf.
41. https://en.m.wikipedia.org/wiki/Reusable_launch_system.
42. www.astronautix.com/craft/stsation.htm.
43. SpaceSecurity.org. 2010. Library and Archives Canada Cataloguing in Publications Data, Space Security. http:/spacesecurityindex.org/wp-content/uploads/2014/10/space.security.2010.reduced.pdf.
44. Space shuttle detailed block program, www.google.com/search?hl=en&client=firefox-a&rls=org.mozilla:enUS:official&channel=np&q=space+shuttle+tailed+block+diagram&bav=on.2,or.r_gc.r_pw.&um=1&ie=UTF8&tbm=isch&source=og&sa=N&tab=wi&biw=1366&bih=624.
45. Private space missions a success, A.M. New York, December 9, 2010.
46. International Space Station Science research accomplishments during the assembly years: An analysis of results from 2000 to 2008. Cynthia A. Evans and Julie A. Robinson, Office of the International Space Station Program Scientist, NASA Johnson Space Center, Houston, Texas; Judy Tate-Brown, Tracy Thumm, and Jessica Crespo-Richey, Engineering and Science Contract Group, Houston, Texas; David Baumann and Jennifer Rhatigan, NASA Johnson Space Center, Houston, Texas, June 2009.
47. www.nss.org/legislative/positions/NSS_Position_Paper_Next_Generation_Space_Stations_2015.pdf.
48. www.nsr.com/news-resources/the-bottom-line/mass-challenge-for-cubesats/.
49. Small satellites are categorized as picosatellites (under 1 kilogram [kg]), nanosatellites (1–10 kg), and microsatellites (10–100 kg). CubeSats, which generally weigh less than 1.3 kg, are in the shape of a cube that can fit in a person's

hand. In the last 50 years, 38 picosatellites, 680 nanosatellites, and 860 microsatellites have been launched worldwide. Henry, H. and S.W. Janson. eds., *Small Satellites: Past, Present and Future*. El Segundo, CA: Aerospace Press, 2009.

50. Prateep Basu, *'Big Data' Leap in EO Markets*, Northern Sky Research, October 1, 2016, www.nsr.com/news-resources/the-bottom-line/big-data-leap-in-eo-markets/.
51. www.russianspaceweb.com/russia_2010s.html.
52. www.spacefoundation.org/programs/public-policy-and-government-affairs/introduction-space/global-space-programs.
53. Orbital paths of U.S., China set to diverge, Jeremy Page, The Wall Street Journal, October 29, 2010.
54. https://en.wikipedia.org/wiki/Chinese_space_program.
55. Japan's Current and Future Programs in Space Exploration 31 January 2017 54th Scientific and Technical Sub-Committee of COPUOS Vienna Masazumi Miyake Director, International Relations and Research Department Japan Aerospace Exploration Agency.
56. https://books.google.com/books?id=u4nXqDvgGrIC&pg=PA303.
57. www.nasa.gov/pdf/296751main_GES_framework.pdf.

3

Satellites

3.1 Fundamental Concepts

A satellite is any smaller object traveling around a larger object.[1] By this definition, the Moon is a satellite to the Earth, and the Earth is a satellite to the Sun. For the purpose of our discussion here, a satellite is meant to be a human-made spacecraft placed in space to orbit another body or some may refer to it as artificial satellite. Some spacecrafts are crewed, such as the space shuttle, but most are un-crewed, such as National Aeronautics and Space Administration (NASA)'s Hubble Space Telescope. If the satellite has an onboard radio transmitter or other energy signals, they are considered as active satellites, and if they do not and only reflect signals that are beamed at them from Earth, they are considered as passive satellites.

After introducing some basic concepts about the satellites in general, the focus of this chapter will be on the application satellites covering remote sensing, meteorological, Global Positioning System (GPS), scientific, research, and communication satellites. Remote sensing and communication satellites will be discussed in more detail due to their important roles that they play in our everyday lives, as it becomes clear when we look into their technologies and applications in different sections of this chapter.

Scientific satellites, also referred to as space science satellites, as their name implies are satellites that carry instruments in order to study an object or a phenomena. Measuring magnetic fields; studying the Sun; or examining the universe in different parts of the electromagnetic spectrum, such as gamma rays, X-rays, ultraviolet light, visible light, infrared light, microwaves, and radio waves are some of these areas of study.

Application satellites survey the Earth's resources whether on the ground, beneath the ground, or inside water bodies. They could also provide weather information to forecasters.

Communication satellites that are considered to be relays in space, relay telephone calls and television signals or data down to Earth. More details on each of the above types of satellites will be provided in this chapter.

Satellites have become an inseparable part of our lives that are so integrated in our daily activities that we sometimes lose sight of their existence and their important roles in our lives. We watch television programs that are distributed to larger geographical zones by satellites; we know how the weather will be in the coming days based on the information collected and analyzed by satellite; we communicate globally using a single satellite or constellation of low-orbit satellites; or we can detect the soil moisture, snow coverage, classify crops, and so forth, using remotely sensed data collected by Earth resource satellites or we use GPS satellites to navigate.

One of the questions that might come up for everyone studying satellites is how many satellites are orbiting the Earth?

According to the Index of Objects Launched into Outer Space maintained by United Nations Office for Outer Space Affairs (UNOOSA),[2] there were 4,256 satellites in August 2016 orbiting the planet, an increase of 4.39% compared to this time in 2015.

From this figure, 221 satellites were launched in 2015, the second highest number in a single year, although it is below the record of 240 launched in 2014. The increase in satellites orbiting the Earth is less than the number launched in the previous year, because satellites only have limited lifespans. The large communication satellites have expected lifetimes of 15–20 years, whereas the small satellites, such as CubeSat's, have much smaller lifetimes of only 3–6 months.

The Union of Concerned Scientists (UCS)[3] details which of those orbiting satellites are operational. According to their June 2016 update, there were only 1,419 operational satellites—only about one-third of the number in orbit. This means there is quite a lot of useless metal hurtling around the planet! This is why there is a lot of interest from companies looking at how they capture and reclaim space debris, with methods such as space nets, slingshots, or solar sails. According to the UCS data, the main purposes for the operational satellites and their numbers are:

- Communications with 713 satellites,
- Earth observation/science with 374 satellites,
- Technology Demonstration/Development with 160 satellites,
- Navigation and Global Position with 105 satellites; and
- Space Science with 67 satellites.

It is noticed that as, it was mentioned above, the majority of the satellites are communication and remote-sensing satellites; 1,087/1,419 or 77%.

Having hundreds of satellites decommissioned after their useful lives are over, the challenge remains in cleaning up the space from the resulting debris crowding and endangering the space activities. Switzerland decided to take action and hence created "janitor satellite" to clean up space and scientists in Lausanne devise Clean Space One to sweep up debris orbiting Earth.[4] Debris include but not limited to bits of long-dead satellites and spent rocket stages that orbit the planet at almost 18,000 mph, each chunk a potential hazard to working satellites or astronauts.

The Switzerland-based Spaceflight Company is finalizing plans with Canada over a potential launch site for a new Private Space Plane, which is to launch a satellite to clean up space junk by 2018. NASA on the other hand, keeps track of 16,000 pieces of orbiting junk that are larger than 10 cm (4 in.) in diameter. There could be more than 500,000 measuring 1–10 cm and many hundreds of millions of smaller ones. One incident creating debris in space was the U.S. satellite Iridium-33 that exploded in February 2009, when it accidentally hit Russia's long-abandoned Cosmos-2251 satellite.

Looking at the statistics of satellites from a different point of view, it is interesting to note that as of December 31, 2015, there were 1,381 operational satellites, serving different functions such as the followings; see Figure 3.1 below.

- Commercial Communications – 37%
- Civil/Military Communications – 14%
- Earth Observation Services (remote sensing) – 14%
- Research and Development – 12%
- Military Surveillance – 8%
- Navigation – 7%
- Scientific – 5%
- Meteorology – 3%

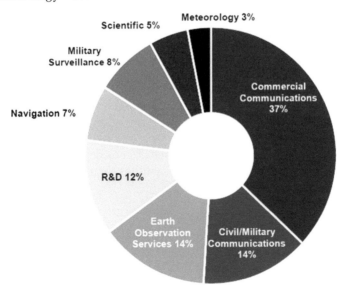

FIGURE 3.1
Operational satellites by function 2015.

Source: **Satellite Industry Association, 2016.**

Again 65% of these satellites are communications and remote sensing and the remaining are other types of satellites.

The number of operational satellites, as detected at the end of 2015, has marked a 39% increase over 5 years, compared to 986 operational satellites reported in 2011. This notable increase is connected to a number of reasons: the average number of satellites launched per year in the 2011–2015 time range has increased by 36% over the previous 5 years period, with small and very small satellites as main contributors to this growth, particularly regarding Low Earth Orbit (LEO) deployments; moreover, the average operational lifespan of certain satellite types, such as GEO communications satellites is expanding. There are now 59 countries with operators represented by at least one satellite, even if some are part of regional consortia.[5]

3.1.1 Types of Satellites

At the time of the writing of this chapter in October 2017, it is interesting to note that two very important space-related milestones took place in this month in history. Seventy-two years ago, the concept of Geostationary Earth Orbit was discussed by Arthur C. Clarke on October 1945 in the *Wireless World (see his article at the end of this chapter),* what later became a famous article, which foresaw the coming of communications satellites in GEO. In his *Wireless World* article, Clarke suggested that three satellites placed in the equatorial plane orbit at an altitude of 36,000 km, spaced 120° apart, could provide global communications. The altitude is specifically important as here satellites rotate at the same angular velocity as that of the Earth, and in the same direction of the Earth's rotation, therefore, appearing to be fixed at the same point above its surface, and hence geostationary. Moreover, it was exactly 60 years ago since the day that a basket ball-sized Sputnik satellite was launched by the Soviet Union on October 4, 1957. Some experts in the field believe that Sputnik was a starting point for the development of the Internet, as we know it today.

These and many more satellite-related concepts will be discussed later in this chapter.

This section briefly mentions the types of commercial satellites and will leave the detailed description of these satellites, in a different order, for the following sections.

- Weather satellites (TIROS, COSMOS)
- Communications satellites (Telstar, Intelsat)
- Broadcast satellites (Direct TV)
- Scientific satellites (Hubble telescope)
- Navigational satellites (GPS Navstar)
- Earth observation satellites (LANDSAT)

Weather Satellite is a type of satellite that is primarily used to monitor the weather and climate of the Earth. Satellites can be polar orbiting, covering the entire Earth asynchronously, or geostationary, hovering over the same spot on the equator.[6] Meteorological satellites see more than clouds and cloud systems. City lights, fires, effects of pollution, auroras, sand and dust storms, snow cover, ice mapping, boundaries of ocean currents, energy flows, etc. Other types of environmental information are collected using weather satellites. Weather satellite images helped in monitoring the volcanic ash cloud from Mount St. Helens and activity from other volcanoes such as Mount Etna.[7] Smoke from fires in the western United States such as Colorado and Utah have also been monitored.

The first weather satellite to be considered a success was TIROS-1, launched by NASA on April 1, 1960.[8] TIROS operated for 78 days and proved to be much more successful than Vanguard 2. TIROS paved the way for the Nimbus program, whose technology and findings are the heritage of most of the Earth-observing satellites that NASA and National Oceanic and Atmospheric Administration (NOAA) have launched since then. Beginning with the Nimbus 3 satellite in 1969, temperature information through the tropospheric column began to be retrieved by satellites from the eastern Atlantic and most of the Pacific Ocean, which led to significant improvements to weather forecasts.[9]

The ESA and NOAA polar orbiting satellites followed suit from the late 1960s onward. Geostationary satellites followed, beginning with the ATS and SMS series in the late 1960s and early 1970s, then continuing with the Geostationary Operational Environmental Satellites (GOES) series from the 1970s onward. Polar-orbiting satellites such as QuikScat and TRMM began to relay wind information near the ocean's surface starting in the late 1970s, with microwave imagery which resembled radar displays, which significantly improved the diagnoses of tropical cyclone strength, intensification, and location during the 2000s and 2010s.

Communication Satellite is a satellite that provides voice, video, and data services. In a more general term, a communications satellite is an artificial satellite that relays and amplifies radio communications signals via a transponder; it creates a communication channel between a source transmitter and a receiver at different locations on Earth. Communications satellites are used for television, telephone, radio, Internet, and military applications. There are over 2,000 communications satellites in Earth's orbit, used by both private and government organizations.[10]

Wireless communication uses electromagnetic waves to carry signals. These waves require line-of-sight, and are thus obstructed by the curvature of the Earth. The purpose of communication satellites is to relay the signal around the curvature of the Earth allowing communication between widely separated points.[11] Communications satellites use a wide range of radio and microwave frequencies. To avoid signal interference, international organizations have regulations for which frequency ranges or "bands" certain organizations are allowed to use. This allocation of bands minimizes the risk of signal interference.[12]

More about such regulations at the international level by ITU (International Telecommunications Union) which is the specialized agency of the United Nations (UN) and other domestic and regional regulatory organizations will be discussed later in this chapter.

Broadcast Satellite or otherwise known as **Satellite Television** is a service that delivers television programming to viewers by relaying it from a communications satellite orbiting the Earth directly to the viewer's location.[13] The signals are received via an outdoor parabolic antenna usually referred to as a satellite dish and a low-noise block downconverter, described in more details in the coming sections of this chapter.

A satellite receiver then decodes the desired television program for viewing on a television set. Receivers can be external set-top boxes, or a built-in television tuner. Satellite television provides a wide range of channels and services. It is the only television made available in many remote geographic areas without terrestrial television or cable television service.

Signals are relayed from a communications satellite on the K_u band frequencies (12–18 GHz) requiring only a small dish less than a meter in diameter.[14] The first satellite TV systems were an obsolete type now known as television receive-only (TVRO). These systems received weaker analog signals transmitted in the C-band (4–8 GHz) from FSS (Fixed Satellite System)-type satellites, requiring the use of large 2–3 m dishes. Consequently, these systems were nicknamed "big dish" systems, and were more expensive and less popular.[15]

Early systems used analog signals, but recent ones use digital signals which allow transmission of the high-definition television signals, due to the significantly improved spectral efficiency of digital broadcasting.

Scientific Satellites. A scientific satellite carries instruments to obtain data on magnetic fields, space radiation, Earth and its atmosphere, the Sun or other stars, planets and their moons, and other astronomical objects and phenomena.[16] More on scientific satellite will be covered later on in this chapter.

Navigational Satellites or Satellite Navigation system is a system that uses satellites to provide autonomous geo-spatial positioning. It allows small electronic receivers to determine their location (longitude, latitude, and altitude/elevation) to high precision (within a few meters) using time signals transmitted along a line of sight by radio from satellites. The system can be used for providing position, navigation, or for tracking the position of something fitted with a receiver (satellite tracking). The signals also allow the electronic receiver to calculate the current local time to high precision, which allows time synchronization. Satellite navigation systems operate independently of any telephonic or Internet reception, though these technologies can enhance the usefulness of the positioning information generated. A satellite navigation system with global coverage may be termed a **global navigation satellite system** (**GNSS**). As of December 2016, only the United States'

GPS, Russia's GLONASS and the European Union's Galileo were global operational GNSSs. The European Union's Galileo GNSS is scheduled to be fully operational by 2020.[17] China is in the process of expanding its regional BeiDou Navigation Satellite System into the global BeiDou-2 GNSS by 2020.[18] India, France, and Japan are in the process of developing regional navigation and augmentation systems as well.

Earth Observation Satellites are satellites specifically designed for Earth observation from orbit, similar to spy satellites but intended for non-military uses such as environmental monitoring, meteorology, map making, etc.

Most Earth observation satellites carry instruments that should be operated at a relatively low altitude. Altitudes below 500–600 km are in general avoided, though, because of the significant air-drag at such low altitudes making frequent orbit reboost maneuvers necessary. The Earth observation satellites Earth Resource Satellite (ERS)-1, ERS-2, and Envisat of European Space Agency (ESA) as well as the MetOp spacecraft of EUMETSAT are all operated at altitudes of about 800 km. The Proba-1, Proba-2, and SMOS spacecraft of ESA are observing the Earth from an altitude of about 700 km. The Earth observation satellites of UAE, DubaiSat-1, and DubaiSat-2 are also placed in LEOs and providing satellite imagery of various parts of the Earth.[19,20] Earth Resource Satellites also known as Remote-Sensing Satellites are yet another term used for Earth Observation Satellites. The first such satellite was Landsat satellites which started in 1972. These satellites and their applications will be explored in much details later in this chapter.

3.1.2 The Benefits of Satellites

Satellites have developed into so many different application areas and disciplines since their inception in the 1960s that it makes it, if not impossible, very difficult to cover all these areas and disciplines. The author will do his best to cover the main and most impacting systems operational today or planned for the future.

During the last five decades, satellites have not only provided information on Earth's resources and the atmosphere surrounding Earth, but also have been instrumental in providing communications to remote, isolated, and deprived locations worldwide. Using satellites for medical purposes, telemedicine; for training and educating employees and students in remote areas, e-learning, distance learning, or distance education; and providing farmers with useful information to optimize their products, are only a few of such benefits. Scientific satellites have broadened human knowledge of the universe; they have revealed planet-wide atmospheric storms on Saturn, the birth of stars in a nearby galaxy, the evolution of supernova, and phenomena that suggest the existence of black holes. Satellites provide a wide coverage of the area under study whether it is a communication satellite that could cover the entire Earth from the geostationary orbit using only three satellites, or a remote-sensing satellite that provides high-resolution images from LEO.

Satellite communications have distinct benefits over terrestrial alternatives by being universal, versatile, reliable, seamless, fast expandable, and flexible. Research and scientific satellites have been helping scientists find answers to the unknown or assisting tourists in finding their way using GPS navigators. Satellites can help cargo ships or passenger ships to navigate in the best possible route taking advantage of the real-time data provided to these vessels. Satellite television provides TV signals in places that are extremely difficult, challenging, or expensive to provide such services by other means. Digital sound broadcasting by satellites provides hundreds of radio channels to users who are traveling or their jobs require them to be on the road most of the time—these are the best quality radio signals that do not fade away or change channels as they move from one area to another. Remote-sensing satellites provide high-resolution images, as was pointed out earlier, from which a vast area of applications are made available. Commercially they can provide soil moisture information on a larger scale for agricultural purposes, can predict the amount of water produced from a volume of snow on the ground when melted, can detect diseased vegetation or trees, and hundreds of other such useful information. Noncommercial remote-sensing satellites can assist with the optimum location of launching a missile from a submarine by providing the thinnest layer of ice cover in places where water bodies are covered by ice, and other such applications.

3.1.3 Satellite Orbits

Each satellite moves within a predetermined path in space above the Earth's atmosphere which is called orbit. Because the air in the atmosphere causes the satellite to slow down, satellite orbits are at least 180 miles (300 km) above sea level. Satellites are launched in a variety of orbits, depending on the satellite's purpose. Most of the satellites are launched in one of three major orbits. Geosynchronous orbit (GEO), which is 22,300 miles (35,888 km) above the equator, has the characteristics that the satellite's velocity and direction of movement matches that of a point on the Earth's equator and because the velocity and direction of movement of satellites in this orbit are the same as that of the Earth's rotation, the ground-based antennas do not need tracking devices. The second type of orbit also frequently used is the medium Earth orbit (MEO) that is between the LEO and GEO. Satellite systems such as GPS satellites are located in this orbit. The third frequently used orbit in the LEO, which could be anywhere from about 200 miles (320 km) to about a 1,000 km (620 miles) above the Earth's equator. This is the level where the space shuttle and the Russian Mir space station are located, as was discussed previously. (More on these orbits will be discussed later in this chapter when the applications of satellites are covered.)

Satellites stay in orbit due to the balance of two effects: velocity, or the speed at which it would travel in a straight line, and the gravitational pull between the Earth and the satellite.

A satellite in orbit moves faster when it is close to the planet or other body that it orbits, and slower when it is farther away from the planet. When a satellite falls from high altitude to lower altitude, it gains speed, and when it rises from low altitude to higher altitude, it loses speed. A satellite in circular orbit has a constant speed that depends only on the mass of the planet and the distance between the satellite and the center of the planet as will be discussed later in this chapter. Figure 3.2 shows the speed in circular Earth orbit, and Table 3.1 shows the other characteristics of these satellites.

The most distant point of a satellite's orbit from Earth is called its apogee. Its closest point to Earth is its perigee. The difference between the perigee and apogee shows the degree of eccentricity of the orbit. Figure 3.3 shows these parameters, and Figure 3.4 shows the different orbits.

3.1.4 Mission Analysis

The primary task in the satellite mission analysis is to select the optimum orbit that best enables the satellite and payload to perform their missions. Providing the mission is feasible, trade-offs are performed in order to find a suitable orbit that meets the mission goals. To calculate the approximate orbits for artificial satellites, Kepler's Laws are used. These laws describe the motion of any two bodies around each other. In all cases, both bodies orbit around the common center of mass, the barycenter, with neither one having their center of mass exactly at one focus of an ellipse. However, both orbits are ellipses with one focus at the barycenter. Satellites that orbit the Earth also follow Kepler's laws that apply to the motion of the planets around the Sun.

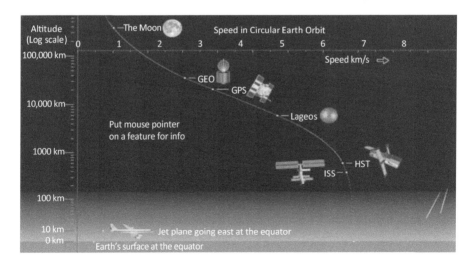

FIGURE 3.2
Speed in circular Earth orbit. (Root, Jeff, 2004, Orbital Speed, www.freemars.org/jeff/speed/index.htm.[21])

TABLE 3.1

Other Characteristics of Satellites

	Altitude	*r*	Speed	Period	Lifetime
The Moon	385,000 km	391,370 km	1.01 km/s	27.3 days	Billions of years
	100,000 km	106,370 km	1.94 km/s	4 days	Billions of years
GEO	35,800 km	42,170 km	3.07 km/s	1 day	Millions of years
Navstar	20,200 km	26,570 km	3.87 km/s	12 hours	Millions of years
	10,000 km	16,370 km	4.93 km/s	5.8 hours	Millions of years
Lageos	5,900 km	12,270 km	5.70 km/s	3.8 hours	Millions of years
	2,000 km	8,370 km	6.90 km/s	2.1 hours	Millenia
	1,000 km	7,370 km	7.35 km/s	105 minutes	Millenia
Hubble	600 km	6,970 km	7.56 km/s	97 minutes	Decades
ISS	380 km	6,750 km	7.68 km/s	92 minutes	Years
	200 km	6,570 km	7.78 km/s	89 minutes	Days or weeks
	100 km	6,470 km	7.84 km/s	87 minutes	Minutes
Sea level	0 km	6,370 km	7.90 km/s	84 minutes	Seconds

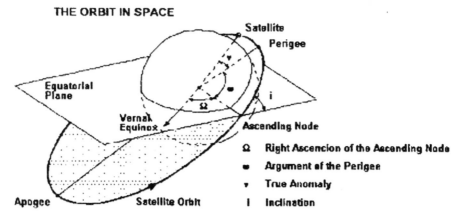

FIGURE 3.3
Apogee and perigee.[22]

Johannes Kepler (1571–1630) derived three laws describing planetary motion which generally apply to any two bodies in space that interact through gravitational force. The more massive of the two bodies is referred to as the primary, and the other is called the secondary or satellite. The three laws of motion derived by Kepler are as follows:

Kepler's First Law
   This law states that "the path followed by a satellite around the Primary will be an Ellipse."[23] Figure 3.5 shows the ellipse with two focal points F_1 and F_2. The center of mass coincides with the center of

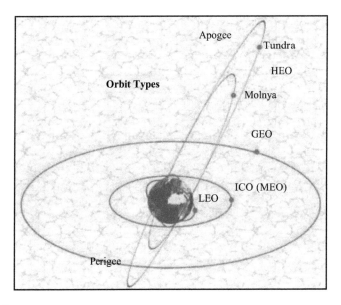

FIGURE 3.4
Different types of orbit.

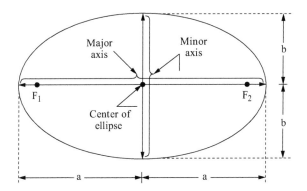

FIGURE 3.5
Kepler's First Law: The foci F_1 and F_2, the semimajor axis a, and the semiminor axis b of an ellipse.

the Earth, the more massive of the two bodies (Earth and satellite), which is at one of the foci. The eccentricity, e, is expressed in terms of the semi-major axis, a, and semi-minor axis, b, of the ellipse by the following equations:

$$e = \sqrt{\frac{a^2 - b^2}{a}}$$

(3.1)

The eccentricity, e, falls within the range of $0 < e < 1$. When $e = 0$, that is when $a = b$, the orbit becomes circular, and when $e = 1$, that is when $b = 0$, the path becomes a straight line. Therefore, the eccentricity gives us the shape of the ellipse between these two extreme limits.

Kepler's Second Law

Kepler's Second Law states that "for equal time intervals, a satellite sweeps out equal area in the orbital plane, focused at the barycenter." Figure 3.6 shows this law graphically. Assuming the satellite travels distances S_1 and S_2 meters in 1 second, then the areas A_1 and A_2 will be equal. The average velocity in each case is S_1 and S_2 meters per second, respectively, and because of the equal area law, the velocity at S_2 is less than that at S_1. This could also be seen because the satellite takes longer to travel a given distance when it is farther away from the Earth because of weaker gravitational force.

Kepler's Third Law

Kepler's Third Law states that "the square of the periodic time of orbits is proportional to the cube of the mean distance between the two bodies." Kepler's third law for artificial satellites orbiting the earth is written as:

$$a^3 = \frac{\mu}{n^2} \tag{3.2}$$

where n is the mean motion of the satellite in radians per second, and m is the Earth's geocentric gravitational constant which is $\mu = 3.986 \times 10^{14} \, \text{m}^3/\text{s}^2$. Equation 3.2 applies when there are no perturbing forces acting on the satellite.

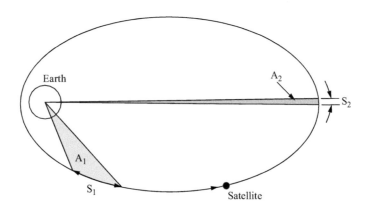

FIGURE 3.6
Kepler's Second Law: The areas A_1 and A_2 swept out in unit time are equal.

3.1.5 Satellite Software

Satellite industry enjoys from number of software developed by many organizations, some free for download and some come with a fee.

The followings are some of these softwares that are instrumental in teaching satellite courses in academia or other training institutions.

Each one of these software will be briefly described and to explore them, references will be given for the instructors and students to enjoy the full downloads and examination of these software.

Having taught satellite communication course for three decades, I would consider the use of these software packages in the classroom environment a very useful tool for visually learning some difficult-to-understand concepts and principles, to supplement the laboratory-based satellite experiments.

1. **SatSoft:** Satellite Antenna Design Software (www.satcom.co.uk/article.asp?article=27)

   SATSOFT is used for communication satellite antenna design, analysis, and coverage planning. Quickly assess antenna coverage and gain, conduct antenna trade studies, develop shaped beam and multibeam antenna designs, and perform other satellite payload engineering tasks.

   www.satsoft.com provides more information on this software. SATSOFT was written for the systems engineer as well as the antenna specialist. Its graphical user interface (GUI) was designed to enable users with even a basic knowledge of antennas to use the software productively. Advanced tools will appeal to the antenna specialist. SATSOFT provides essential tools for planning GEO, MEO, or LEO satellite communications system. It has three basic types of views: Composite View has a hierarchical tree in its left pane showing all objects that are currently modeled. The right pane is a map of the earth showing antenna contours and polygons for all satellites. Double clicking on an object in the tree opens either an Antenna View, for designing multibeam antennas, or a Polygon View, for designing coverage polygons used in the design of the antenna. Many tools are provided for various types of space-based communications systems analysis and design. Basic features include map and contour plotting, coordinate conversions, file import/export, boresight rotation, city pointing boxes, visibility contours, ground tracks, EIRP tables, and other features. Context-sensitive help can be accessed at any time by clicking on Help from any menu or dialog box.

   SATSOFT Complements Other Software, but is suitable for use by itself or with antenna modeling programs such as GRASP and POS (TICRA Engineering Consultants, Copenhagen, Denmark), or with other modeling software. Although SATSOFT contains a powerful set of analysis and design capabilities, the program is designed to interface to and complement other analysis software. Antenna patterns generated with SATSOFT may be used with Satellite Tool Kit (STK), for example.

Key features of this software are:

- Simplified Antenna Coverage Synthesis create a rectangular or triangular grid of beams over the coverage area, using the Gaussian Beam Model provided with SATSOFT Standard, an optional reflector (SATSOFT/AR and SATSOFT/PO) or array (SATSOFT/PAM) model, or pattern data read from a file. Then create a grid of synthesis stations, and specify the desired gain at each. Combine the component beams into a composite coverage using SATSOFT's mini-max optimizer, which attempts to achieve the target gain at each synthesis station by modifying the beam excitations. A dual-mode optimizer is also available at additional cost. A number of tools are provided to automate the coverage design process. Component beam and synthesis station grids can be created automatically by filling a coverage polygon. A graphical editor is provided to interactively add, delete, move, or re-specify the parameters of single or multiple beams or stations just by pointing and clicking the mouse. What used to take hours can now be done in minutes. What's more, computation is very fast, thanks to 32-bit operation and highly optimized algorithms for drawing maps, contouring patterns, and computing beams. Any number of component beams and synthesis stations can be handled simultaneously. Every satellite may contain as many antennas as desired, and each antenna is a separate object with its own design view. A tabbed workbook interface makes it easy to switch between views.

- Contour Plots of EIRP, G/T, C/I, and Other Parameters. A fast contouring engine within SATSOFT produces plots quickly and accurately. Intelligent auto-ranging and zoom-in/zoom-out functions make it easy to view a desired range of data. You can also create plots at a precise scale for overlaying. Conversion from directivity or gain to other parameters is accomplished by specifying a dB offset to be applied to the antenna pattern prior to contouring. By applying slant-range loss compensation, flux-density contours can be computed. SATSOFT will also plot copol/crosspol ratios for vector antenna patterns.

2. **STK:** Analytical Graphics, Inc. (AGI) (Exton, Pennsylvania) offers commercial off-the-shelf (COTS) analysis and visualization software to more than 28,000 aerospace, defense, and intelligence community professionals through its core product STK and a comprehensive set of STK add-on modules.[24] With integrated land, sea, air, and space elements, the STK software suite technology is ideal for all phases

of complex industry programs from system concept and design to real-time operations. The author had the opportunity to explore the potentials and capabilities of this software package in a 5-day workshop that provided hands-on experience with a variety of the features and functions on which STK is built. AGI's GUI introduced various land, sea, air, and space objects, along with their associated properties and tools that are part of the STK software suite. The course demonstrated how to accurately model, analyze, assess, and plan all aspects of a mission, including orbit/trajectory design, attitude simulations, communication platforms, and global and custom ground and object coverage.

Some of the STK capabilities in the analysis of land, sea, air, and space are briefly outlined as follows[25]:

Conjunction Analysis

   Daily space object catalog updates

   Close approach determination

   Probability of collision

   On-orbit operation

   Laser clearinghouse

   Launch window analysis

Communications Link Analysis

   Single and multihop links

   Analytical and user-defined power spectral density and filters

   Adjacent channel interference

   Light speed signal propagation delay modeling

   Terrestrial propagation

   Interference and jamming

   Laser communication models

   Collection of built-in transmitter and receiver models

   User-defined models for transmitters, receivers, antennas, and
      phased arrays

Spacecraft Design and Operation

   Solar array power analysis

   Spacecraft exposure to ionizing particles and thermal radiation

   Mission plan to real-time data comparison

   Attitude modeling

   System Performance Analysis

   Lunar and Interplanetary Analysis

   Orbit Determination

STK is provided to the educational establishment by AGI through a signed contract at no charge, and a great number of universities are taking advantage of this very strong, solid, comprehensive, and detail-oriented software for their educational purposes. For more information, the reader is referred to www.agi.com/home.

3. **SMW:** SMWlink is a useful Satellite Link Budget Calculator for the Satellite Communication Professional. From System Noise Figure to G/T, Swedish Microwave AB's SMW link makes the everyday link budget analysis easier for Satellite Communications Professionals. In this user-friendly software, the user visually observes the affect of factors such as rain, antenna elevation angle, frequency band, etc. on the G/T (commonly refered to as Figure of Merit), and the quality of the picture.

     https://smw.se/smwlink/

4. **SatSoft:** SATSOFT speeds the process of planning, designing, and marketing communication satellite payloads. Quickly assess antenna coverage and gain, conduct antenna trade studies, develop shaped beam and multibeam antenna designs and complete many other tasks required for payload design and regulatory filings.

     www.ticra.com/products/software/satsoft

     Its GUI enables users with even a basic knowledge of antennas to use the software productively. Advanced tools will appeal to the antenna specialist.

     - The tool of choice for communication satellite antenna design, analysis, and coverage planning.

     - Enables you to quickly assess antenna coverage and gain, conduct antenna trade-off studies, develop shaped beam and multibeam antenna designs, and perform other satellite payload engineering tasks.

     - Written for the systems engineer as well as the antenna specialist.

     - Provides users with just a basic knowledge of antennas a tool that can be used very productively.

     - Advanced tools will appeal to the antenna specialist.

     - If you need software to model satellite antennas, for GEO, LEO, or MEO constellations, SATSOFT was written for you.

5. **SatMaster:**

   Overview

     SatMaster Pro, is a tool mainly used for preparing satellite link budgets and predicting sun outage events. A demo version, with some restrictions, is available to download for evaluation.

     www.satmaster.com/linkbudget.html

Digital Link Budget Calculators

Full up-down link budget calculators with input provision for uplink power control (UPC), automatic level control (ALC), various interferences and multiple carriers. Finds high power amplifier (HPA) size, uplink power requirements, bandwidth, and power usage per carrier. Calculates atmospheric losses and rain fade margins for any desired availability. Handles 1 bit/symbol (BPSK), QPSK, M-PSK/M-APSK, or M-QAM with modcod (Adaptive Coding and Modulation, a technology that can automatically change the modulation and forward error correction of a link) selectors for DVBS, DVBS2, DVBS2-X, and NS3. Determines power or bandwidth limited modes and calculates the power equivalent bandwidth, if appropriate. Supports paired carrier, turbo and manual modcods.

- Modcod selectors for NS3, NS4, DVBS, and DVBS2 (including DVBS2-X extensions).
- Option to optimize the split of total availability between uplink and downlink.
- Ability to translate excess margins to increased availability.
- Option to calculate carrier EIRP (beam peak) and PSD (beam peak) values.
- ITU-R recommendations P.618-12, P.676-11, P.836-5, P.839-4, P.840-6 supported.
- Choice between C/I or C_{sat}/I_o interference inputs.
- Option to force clear sky calculations for uplink and/or downlink for ACM or aircraft.
- Support for Crane rain models.
- Calculates atmospheric absorption, rain/cloud attenuation, and tropospheric scintillation.
- Supports P, L, S, C, X, K_u, K_a, and K band.
- Produces link budgets for LEO, MEO, HEO, and GEO satellites.
- Support for UPC, ALC, and spread spectrum.
- Modules for 1/2 link budgets for regenerative transponders or quick TVRO antenna sizing.
- Separate bandwidth calculator for NS3, NS4. DVBS, DVBS2, and DVBS2-X.
- Sun outage prediction (single site and satellite).
- Sun outage batch file handling (multiple sites or multiple satellites).

- Antenna aiming.
- Dual/multi feed positioning.
- Dish sizing.
- Polar mount alignment.
- Automatic altitude and magnetic variation calculation.
- Solar transit times.
- Off-axis gains.
- Loads and displays decoded two line element (TLE) files (SGP4/SDP4 model).
- Module for batch processing of up to 10,000 GEO link budgets
- Ionospheric scintillation estimator.
- Optional four satellite, ASI calculator within GEO 'bent pipe' link budget modules.
- Automatic antenna noise calculation option.
- Tabulation of R0.01 values and rain attenuation for a selected country.
- Key link budget results are highlighted.
- Non-reflector gains and effective apertures accepted
- Generates numerous graphs and tables.
- Numerous satellite-based calculation tools such as G/T, off axis antenna gains and bandwidth.

6. **Link Budget Calculator**
   a. www.satsig.net/linkbugt.htm
      This web site is to promote legitimate, satcom access for people in all locations, who are unable to gain access using terrestrial ADSL via copper or optical fiber phone lines or using cable modems. Satcom is an alternative and provides independent small-dish two-way access from anywhere except the extreme polar regions. There are many features making this software a very strong tool for students to learn different satellite service parameters by inputting some of the link parameters such as uplink frequency in GHz, Uplink antenna diameter in meters, uplink antenna aperture efficiency, uplink antenna, or power at the feed in watts, uplink antenna, power at the feed and calculate the remaining parameters.

7. **Sat-Coord:** Sat-Coord is a modular software suite, which supports the processing of satellite network information filed with the ITU, intersystem interference calculation (including $\Delta T/T$ and C/I), IFIC processing and frequency coordination support.

The software has undergone significant testing and development over a period of more than 12 years and has been used extensively to support the satellite coordination activities of RPC Telecom's clients including Intelsat, YahSat, VINASAT, THURAYA, TONGASAT, SingTel, HELLAS-SAT, SUPARCO, ETISALAT, SES Americom, ICO, Hughes Network Systems, O3B, JRANSA, DirecTV, the Cyprus Ministry of Communications, the Nigerian Communications Commission, INDOSAT, Es'hailSat, ANGKASA, Paradigm, BRIsat, KACST, and the Government of Australia.

Sat-Coord can be downloaded and registered for a free, fully featured, 30-day trial.

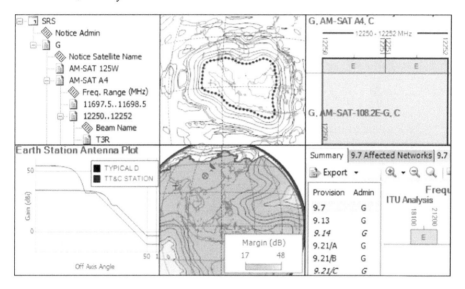

8. Satellite Tracking

J-Track 3D Satellite Tracking

This is a good software developed by NASA and was offered to users up until recently. It provided the real time tracking of LEO, MEO, and GEO satellites and had many interesting features. It might be made available soon.

Some of the other satellite software that are used are briefly mentioned below: www.satobs.org/orbsoft.html

- **Quicksat**

  This program uses two-line elements to generate predictions which include an estimate of magnitude. It is text based, reading the input parameters from a formatted file. A comprehensive magnitudes file is packaged with it.

- **IDSat**
  IDSat assists in the identification of observed satellites, by computing and tabulating close appulses of known satellites relative to the observed time and position. IDSat also lists the direction of travel, angular velocity, and estimated prediction accuracy. Subject to the availability of data, it can also estimate the predicted visual magnitude.

- **ObsReduce**
  ObsReduce, is an MS Windows program for positional observers, that reduces observations of satellites relative to the background stars, into their precise coordinates. Observers identify their reference stars in a simulated binocular or telescope field of view, enter the observed geometry, and the program automatically produces an observation report in any one of the standard reporting formats.

- **TrakSat 4.09**
  TrakSat shareware program generates predictions from TLEs. It also produces a graphical display of the ground track in various projections and a plot of the pass against the local (star-filled) sky. Uses GUI interface under MSDOS. Propagation models used are SGP4 and SDP4. The Home Page for TrakSat contains information about the latest version of TrakSat and allows a trial version to be downloaded.

- **WinOrbit 3.4**
  This Windows freeware program generates a ground track graphic using two line or AMSAT format elements. Able to manipulate element sets in the case of delayed launches.

- **Sky Chart 2000.0**
  This program is primarily a planetarium program, but it can compute and display artificial Earth satellites using standard TLE satellite files.

- **Home Planet (Release 2 and 3)**
  This program provides a combination astronomy and sat tracking program for Windows.

3.1.6 Satellite Regulations

In order for all citizens and businesses to fully benefit from the services that satellites provide it is necessary that:

- Sufficient Spectrum is available and it is used with the highest efficiency,
- International and national regulations keep pace with evolving satellite services, and
- Licensing regimes must provide flexibility.

To address the first requirement and to use the allocated spectrum to satellites by ITU on a global basis in most efficient way, satellite operators are developing and investing in new satellite networks and technologies that will result in a more robust satellite global infrastructure. It is forecasted that over 2,000 satellites are expected to be launched during the next decade implementing technologies that expand traditional FSS applications such as:

- Steerable High Capacity Beam to allows delivery of broadband satellite services anywhere anytime
- Tracking earth stations to allow mobility operations in the FSS bands to complement those offered in the Mobile satellite service (MSS)
- Flat panel antennas to allow satellite services to be delivered in more locations
- Coding rates to allow increases in availability and delivery of more information per Hz.

These and many other technologies such as improving the satellite antenna technology to allow satellites to be located in the GEO orbit closer without causing interference into the neighboring satellites, taking advantage of frequency reuse, to operate at higher frequency (K_a-Band) to increase the available bandwidth, etc. will result in using the frequency spectrum, which is a naturally limited resource, optimally.

To address the second requirement, Satellite Service Regulations, the following steps need to be taken:

- Satellite service areas to cover relatively large geographic areas while meeting the required international and national regulations,
- Regulations to keep pace with satellite technologies so that consumers may benefit from new satellite services,
- To follow the International framework—ITU Radio Regulations (RR) treaty that regulates access to the orbit and spectrum resources—modifications that takes place only at World Radio Conferences (WRCs, occurs every 3–4 years),
- For each country to establish their own national regulatory regimes and harmonize it on a regional level,

On a regional basis, regulations should be coordinated among the three ITU-identified regions.

- In Europe, regulation for satellite services are generally harmonized,
- In the Americas, regulation for satellite services typically vary on a country-by-country basis; there are differences in licensing satellite services by sector, e.g., maritime versus aeronautical and differences with respect to domestic versus foreign terminals.

- CITEL has adopted Recommendations that would foster more robust deployment of satellite services in the Americas, such as:
- PCC.II/REC.52   (XXVII-16)—Licensing   Regimes   for Ubiquitously Deployed Satellite Service Earth Stations
- PCC.II/Rec.50 (XXVII-16)—Authorization of Earth Stations in Motion Communicating with Geostationary Space Stations in the FSS in the Frequency Bands 19.7–20.2 GHz/29.5–30.0 GHz in the Americas

ITU's mission is "to ensure rational, equitable, efficient and economical use of the radio frequency spectrum by all radiocommunication services including those using the geostationary satellite orbit or other satellite orbits – and to carry out studies on radiocommunication matters."

Legal Framework of ITU could be summarized as follows:

a. **UN Outer Space Treaty (1967)**

- Outer space is free for exploitation and use by all states in conformity with international regulations,
- States retain jurisdiction and control over objects they have launched into outer space,

b. **ITU RR**

Part of the ITU Administrative Regulations and Instruments complementing the provisions of the ITU Constitution (CS) and Convention (CV), which govern the use of telecommunications.

- *Legal treaty – bindings* on all Member States
- Principles of use of orbit/spectrum (CS and RR).
- Allocation of frequency bands and services.
- Procedures and Plans.
- Ratification of the ITU Convention (CV) implies acceptance of the ITU RR.

3.1.7 Economics of Satellite Systems

Depending on which satellite system we are discussing, the economics would be different. For example, it costs the same to transmit a television signal via satellite to one or a million receivers.

Figure 3.7 below illustrates that the Rate of Return on Investment stays flat for satellite since it is independent of the population density but for terrestrial systems, this rate drops as the population density is reduced. When the green line crosses the blue line, then it is preferred to use satellite rather than terrestrial. When the acceptable ROI is parallel to the blue line, then satellite will be the technology of our choice for both rural and urban areas.

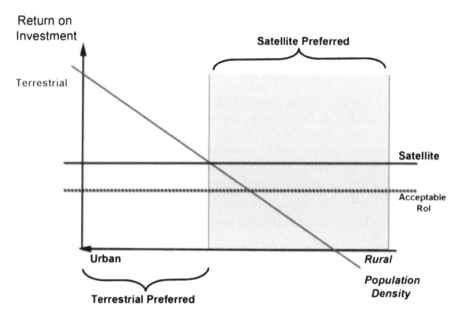

FIGURE 3.7
Rate of Return on investment for satellites and terrestrial systems.

Source: **Satellite Executive Briefing, Volume 7 Number 8, September 2014.**

Satellites are also able to interconnect locations thousands of kilometers apart because satellite is distance insensitive unlike other communication systems. This unique characteristic of satellites implies that the cost of satellite communication link remains the same for short or long distances.

As a result of distance insensitivity of communication satellites, they are particularly better systems for certain applications which terrestrial telecommunications, like microwave relay, coaxial cables, and optical fiber cable are not.

The applications where costs of service and technological advantage go to communications satellites include but not limited to: need for global coverage, access to hard-to-reach areas, rapid installations and commissioning of a communication link, direct broadcast television or large-scale television/radio distribution; long distance overseas telecommunications services where large numbers of countries need to be interconnected (regional systems); mobile communications for aircraft; ships and land vehicles on long distance journeys; data distribution services (e.g., news, electronic publishing, etc.); digital sound broadcasting; and rural and remote communications services.

The economics of satellite communications are also affected by rapid technological innovations and by economies of scale in their operation.[26]

The space sector has distinguishing features such as the use of cutting edge technologies and longer terms for both project development and return on investments. It is a government dominated sector, as access to space is costly, involves technical risks, and the viability of space enabled services requires large users markets. Linkage to defense is deeply rooted since the very beginning of the Space Age, as during the Cold War space activities were also a tool of political and military confrontation between United States and Union of Soviet Socialist Republic (USSR). Nowadays, despite the growing importance of commercial space sector, the use of space for defense purpose remains prominent thanks also to space technologies that have both civil and military applications like weather forecast, which can also be used for early warning, remote sensing, with its applications in intelligence, and global navigation, a precision targeting system.[26]

3.1.7.1 Space Value Chain: A Breakdown

By definition, a value chain is the whole range of activities, including design, production, marketing, logistics, and distribution to support the final customer that organizations engage in to bring a product to the market, from conception to final use. Each step adds some form of value.

The space sector value chain, shown in Figure 3.8 demonstrates how values are added in each part of the chain. We are witnessing lately that space economy is attracting much more attention worldwide, as public and private investors look for new sources of economic growth and innovation, and space economy has become a relevant domain for high-tech innovation,

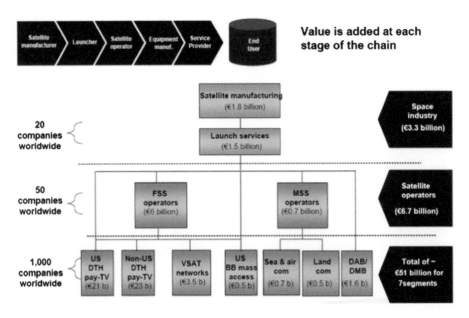

FIGURE 3.8
The space value chain. (Credits: ESOA (European Satellite Operators' Association).)

commercial opportunities, and strategic purposes. As shown in the figure, different kinds of activities, inputs, and processes contribute to shape the global space value chain.

The global space industry has grown steadily, with global space activities amounting at \$314.17 billion in 2013.[27] More than 60% of space-related economic activity comes from commercial goods and services. Commercial space infrastructure and support industries include satellite manufacturing, launch services, ground stations and related equipment, while commercial space products and services encompass satellite broadcasting, communication, Earth observation, geo-location, and global navigation equipment and services. The latter is one of the space industry's fastest-growing sectors.

DTH (Direct to Home) television services represent a majority part of revenues in the commercial space products and services sector. Satellite communications market is composed of satellite operators, which lease out the transmission capacity (transponders) of their satellites to private and government clients in need of transmission capability. According to ITU this sector can be divided in FSS and mobile satellite services (MSSs). FSS refers to the delivery of satellite communications to stationary ground receivers that can be moved from one location to another but do not work while in transit. Capabilities offered by MSS are similar, but the communication link connects with mobile receivers, such as satellite telephones or in-flight communications.

Satellite technology enabling Earth observation and imaging services constitutes a small, but very important part of the global space economy, providing a wide spectrum of applications in different fields; the largest revenue growth occurred in defense, intelligence, and in the sectors of agriculture, energy, and natural resources.

Space systems play an increasing role in the modern societies, their economic development, and their strategic order. Satellite technology applications in variety of areas such as navigation, communications, meteorology, and Earth observation have and continue to play a major role in the economic development of countries. Space technologies also affect agriculture planning, disaster management, medicine, land monitoring, transportation, and urban planning, all the foundations of the economic infrastructure of countries. Satellite's diverse fields of applications make space an engine of economic growth. "Despite the economic crisis, institutional space budget faced less decrease than other productive sectors. OECD underlined that, even though space often has a reputation of being an expensive industry, all G20 Countries invest in their respective national space program a very small percentage of the GDP (Gross Domestic Product). The benefits provided by the use of space assets range from qualitative aspects, such as strategic advantages and improved decision-making processes based on satellite imagery, to monetarily quantifiable ones, such as cost-efficiency and increased safety derived from using satellite navigation tools for ground, maritime and aviation applications. Scientific and technological innovations are making space applications accessible to more people, being vehicle of social and economic growth also for developing Countries."[28]

Global space industry revenues topped $323 billion in 2015, with 76% comprising commercial space products and services and commercial infrastructure. The space industry has also grown by almost 10% between 1998 and 2015—much more than global GDP growth over the same period of time. In this context, the satellite industry has proven to be the dominant driver of growth, accounting for more than 62% of space industry revenues in 2015, the majority of which was generated by satellite services such as telecommunications, Earth-observation, science and national security applications. In the years 2009–2015, satellite applications were dominated by technology (47%), whereas projected trends show that as from 2016, Earth-observation will take the lead with 73% of the applications market.

The average number of satellites launched globally per year increased by 36% in the years 2011–2015 over the previous 5 years, with a number of total operating satellites reaching 1,381 in 2015 as compared to 986 in 2011.

Small satellites are expected to take a relevant stake of the projected industry growth: 28 nano/micro satellites were launched into orbit in 2008, increasing to 141 in 2014, whereas more than 3,000 are expected to be launched between 2016 and 2022. Of particular interest is the rapid adoption of the CubeSat standard of small satellite which is the first globally and academically recognized standard for small satellites with specific weight and volume requirements.

3.2 Application Satellites

A large number of satellites are used for a variety of applications, as briefly described in the previous section. Such applications include remote-sensing satellites, meteorological satellites, and global positioning satellites. Weather satellites are vital in predicting when tropical storms, hurricanes, floods, cyclones, tidal waves, and forest fires may strike. Having this information helps us to prepare for these events and avoid disaster. Remote-sensing satellites, as mentioned earlier, can estimate the amount of water produced from a volume of snow when melted, which again helps to prevent any damage resulting from a possible flood. Crop classification, soil moisture measurements, plant disease detection, plant maturation time for optimum harvest scheduling, and so forth, are only a few of remote sensing satellites' applications. Hundreds of applications of the GPS have been developed, affecting every facet of modern life. GPS has been applied to agriculture, aviation, environment, marine, public safety and disaster relief, recreation, railroads and highways, space, and surveying and mapping. A more detailed discussion will be provided in the remaining of this chapter addressing the above and more application areas.

3.2.1 Remote-Sensing Satellites

By definition, "Remote sensing is the science and art of obtaining information about an object, area, or phenomenon through the analysis of data acquired by devices that is not in contact with the object, area, or phenomenon under investigation."[28] This is how remote sensing is referred to in related literature. Many different instruments can be used to collect remote-sensing data, such as binoculars, telescopes, or cameras. Remote sensing is an important way that we can study many features on or under the Earth's surface without being in contact with the target under study. The three common platforms used for collecting remote-sensing data are ground-truth, airborne, and space-borne. Collecting data using remote-sensing sensors such as radiometers, radar, scatterometers, and so forth, for example, from the boom of a truck is a ground-truth remote-sensing data collection. Aerial photography is considered an airborne remote sensing where the sensors are located on the belly of an aircraft flying at low or high altitude, depending on the area under investigation. Satellites, as the space-borne platform, can also collect remote sensing which can turn into information. Remote-sensing satellites orbit hundreds or even thousands of miles above the Earth. Because most remote-sensing satellites do not use film, the pictures made from the data collected from them are often called images. Satellite images are used to study large areas in less detail. Remote sensing from space has developed in recent decades from an experimental application into a technology that we depend on for many aspects of Earth and planetary research. Satellite remote-sensing systems provide us with data critical to weather prediction, agriculture forecasting, resource exploration, and environmental monitoring.

The author believes that engineers and applied scientists need to appreciate the value of remote-sensing technologies, particularly now that a number of new space-borne systems are becoming operational and the data from them are commercially available. With the launch of systems such as Ikonos and QuickBird that are very high-resolution systems and the new generations of Landsat, Radarsat, Spot, and other optical and microwave remote-sensing satellite systems that are generating a huge database that could be used in numerous applications in a wide variety of disciplines, the need for understanding remote-sensing technology and utilizing the satellite generated data is becoming a necessity for every engineer and technologist. The newest tools for civilian satellite imaging are quite revolutionary, and just emerging at the beginning of the 21st century. These systems typically involve panchromatic imagery with resolutions below 1 m, and color imagery with resolutions as high as 2.5 m, which is better spatial resolution than the traditional remote-sensing satellites such as Landsat and Spot. Many different products and services result from these new sensors.

> The availability of high-resolution imagery brings opportunities for improving world security, the environment and the quality of life for all humanity. If objectively applied, it can be an effective deterrent to international acts of aggression. However, it also brings opportunities for the negative elements of humanity to aid and encourage criminal activities and for commercial espionage, on a global, regional and local scale.[29]

What follows in this section are brief descriptions of some of the more widely used remote-sensing satellite systems that are used worldwide. Information is provided for the high-resolution remote-sensing satellite systems, then followed by the lower-resolution satellite systems that have been in operation for more than three decades.

Remote-sensing satellites operate in different parts of the electromagnetic spectrum. The two most commonly used spectral bands by remote-sensing satellites are microwave and optical. By definition, the electromagnetic spectrum is the range of all possible frequencies of electromagnetic radiation. The electromagnetic spectrum of an object is the characteristic distribution of electromagnetic radiation emitted or absorbed by that particular object.

The electromagnetic spectrum, as shown in Figure 3.9, extends from low frequencies used for modern radio through to gamma radiation at the short-wavelength end, covering wavelengths from thousands of kilometers down to a fraction of the size of an atom.

Remote-sensing satellites worldwide are shown in Table 3.2 and Figure 3.10 ranging from low-resolution to the highest commercially available resolution satellite, QuickBird, with the understanding that this chart is by no means a

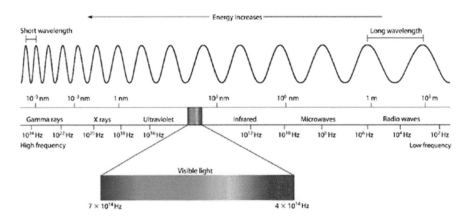

FIGURE 3.9
The electromagnetic spectrum.

TABLE 3.2

Remote-Sensing Satellites Worldwide

System	Operator	Current Satellites	Type	Highest Resolution (m)
EROS	ImageSat International	EROS A	Optical	1.5
		EROS B	Optical	0.7
		EROS C	Optical	0.7
Ikonos	GeoEye	IKONOS-2	Optical	0.8
OrbView	GeoEye	OrbView-2	Optical	1,000
GeoEye	GeoEye	GeoEye-1	Optical	0.41
QuickBird	DigitalGlobe	EarlyBird	Optical	3
		QuickBird-1	Optical	1
		QuickBird	Optical	0.6
Radarsat	MDA	Radarsat-1	Radar	8
		Radarsat-2	Radar	3
SPOT	Spot Image	SPOT 2	Optical	10
		SPOT 4	Optical	10
		SPOT 5	Optical	2.5
WorldView	DigitalGlobe	WorldView-1	Optical	0.5
Disaster Monitoring Constellation	DMC International Imaging	AISAT-1 (Algeria)	Optical	32
		NigeriaSAT-1 (Nigeria)	Optical	32
		UK-DMC (United Kingdom)	Optical	32
		Beijing-1 (China)	Optical	4
TerraSar		TerraSar-X	Radar	1
RapidEye	RapidEye	RapidEye-1	Optical	6
		RapidEye-2	Optical	6
		RapidEye-3	Optical	6
		RapidEye-4	Optical	6
		RapidEye-5	Optical	6

Source: Adapted from www.oneonta.edu/faculty/baumanpr/geosat2/RS-Introduction/ RS-Introduction.html.

comprehensive list of remote-sensing satellites. Some of these satellites will be discussed in this chapter in more detail.

3.2.1.1 Satellite Remote Sensing and Climate Change

Climate change has been the focus of studies and concerns globally for over two decades. In this section, after a quick review of some of the background developments in this area, we will explore the role of satellites and especially the remote-sensing satellites in monitoring and tracking the changes

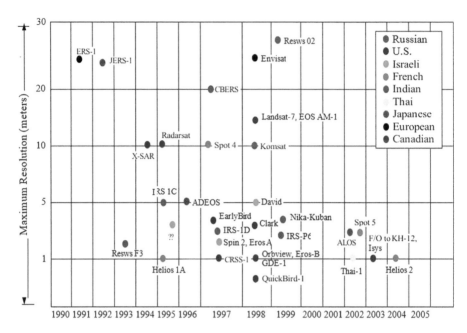

FIGURE 3.10
Remote-sensing satellites and their resolutions: 1990–2005.

in global climate. In 1992, countries joined an international treaty, "the United Nations Framework Convention on Climate Change," as a framework for international cooperation to combat climate change by limiting average global temperature increases and the resulting climate change, and coping with impacts that were, by then, inevitable. By 1995, countries launched negotiations to strengthen the global response to climate change, and, 2 years later, adopted the Kyoto Protocol. The Kyoto Protocol legally binds developed country Parties to emission reduction targets. The Protocol's first commitment period started in 2008 and ended in 2012. The second commitment period began on January 1, 2013 and will end in 2020. There are now 197 Parties to the Convention and 192 Parties to the Kyoto Protocol.

The 2015 Paris Agreement, adopted in Paris on December 12, 2015, marks the latest step in the evolution of the UN climate change regime and builds on the work undertaken under the Convention. The Paris Agreement charts a new course in the global effort to combat climate change.[30]

The Paris Agreement seeks to accelerate and intensify the actions and investment needed for a sustainable low carbon future. Its central aim is to strengthen the global response to the threat of climate change by keeping a global temperature rise this century well below 2°C above pre-industrial levels and to pursue efforts to limit the temperature increase even further to 1.5°C. The Agreement also aims to strengthen the ability of countries to deal with the impacts of climate change.

According to the final report of the expert group on "Towards a European Operational Observing System to Monitor Fossil CO_2 emissions,"[31] specific measurements of atmospheric CO_2 from space are needed for improving fossil CO_2 emissions estimates. This report continues to point out that over the next decade, a succession of partially overlapping missions with a range of CO_2 and CH_4 measurement capabilities will be deployed in LEO. Each mission has been conceived with unique capabilities, designed to improve the measurement precision and accuracy, as well as the spatial and temporal resolution, and coverage to improve understanding of surface fluxes of GHG from the continental to local scales. Current missions are not designed to quantify emissions but natural fluxes. Exceptions being OCO-3, designed to take XCO_2 repeated images over selected hot-spots, and CarbonSat that was designed to take images of XCO_2 (and XCH_4) with 250 km swath, for quantifying emissions when the satellite passes over hotspots.

The Phase 2 of the three-phased Roadmap outlined in the report refers to the use of satellites for the measurement of the CO_2 emissions: "Phase-2 Pre-operational phase – first satellite mission. This phase will be a requirement-driven and user-driven pre-operational phase centered on the development of a first European satellite mission that will deliver the first high resolution satellite data to quantify fossil CO_2 emissions. The data from this mission will be combined with ground-based inventories by a dedicated data assimilation system, to deliver a first set of improved emission estimates at national and subnational scale. International cooperation including data sharing agreements and the definition of optimized plans for the setting up of a global observation system, will enable the assimilation of additional satellite data, e.g. from existing and planned US (OCO-2, OCO-3), French MICROCARB (CNES), Chinese (TanSat) and Japanese (GOSAT, GOSAT-2) missions." In Phase 3 of this roadmap, the operational phase initiated in Phase 2 will be developed using a set of LEO satellites, possibly complemented by Geostationary Earth Orbit (GEO) imagers that will feed into a fossil emission operational data assimilation system. Figure 3.11 below shows that while the carbon dioxide emissions are remaining approximately constant (within a few percent) over the 1990–2010 period, the global emissions that include contributions from non-Annex I countries exhibit a significant increase:

Figure 3.12 below shows the relative contributions of different groups of countries to global fossil CO_2 emission.

Figure 3.13 below demonstrates the recent, current, and planned missions for atmospheric CO_2 column measurements by several satellite systems.

It is to be noted that OCO-2 and GOSAT, GOSAT-2 have pointing capabilities that allow focus on some emitting hotspots if desired, but the prime objective of these missions is to quantify the large-scale distribution of natural fluxes. Also, it is to point out that CarbonSat was not recommended by the science advisory committee of ESA, for the Earth Explorer 8 program. A notional timeline for developing a European contribution to a global

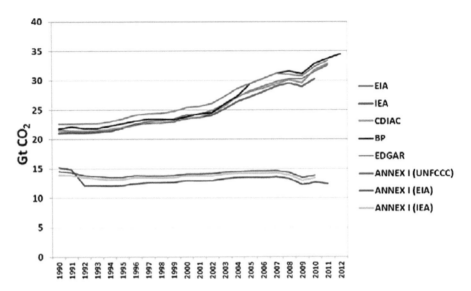

FIGURE 3.11
Global and developed countries (Annex I) carbon dioxide emissions from fossil-fuel use.

Sources: **BP (2013); CDIAC (2013); EIA (2013); IEA (2012); JRC/PBL (2011) UNFCCC (2013). Units = Gt CO_2, i.e., billion tons of CO_2. (UNEP (2013) Appendix 2A. The Emissions Gap Report 2013. United Nations Environment Program (UNEP), Nairobi. © UNEP 2013.**

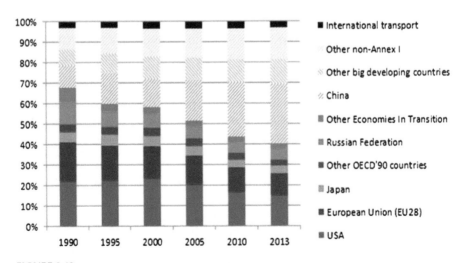

FIGURE 3.12
Relative contributions of different groups of countries to global fossil CO_2 emission. The hatched stacks represent non-Annex I countries. (Graph based on data from EDGAR v4.3 (EC-JRC/PBL, 2015) and Olivier et al. (2014). The period is from 1990 to 2013.)

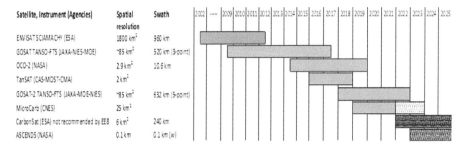

FIGURE 3.13
Recent, current, and planned missions for atmospheric CO_2 column measurements. Green = missions that will advance knowledge of natural fluxes. Red hatching = mission that provide information for separating emissions from natural fluxes.

combination of satellites with capabilities to resolve emissions is given in Figure 3.14 below:

OCO-3 focuses on cities emissions. The other three missions named "CO_2 mission in Copernicus" could be part of a dedicated European Copernicus-CO_2 program.

According to the above referenced report, in parallel to the development of a first European CO_2 satellite mission, the optimal design of the operational fossil CO_2 emission observation system should be assessed through a detailed cost/benefit and system engineering analysis. Notionally, the system should be based on three development phases as outlined in Figure 3.15.

The report emphasizes how important satellite based platforms are in monitoring the CO_2 emissions worldwide and especially remote-sensing technology and remote-sensing satellites. Remotely sensed observations constitute a crucial component since it provides repeatedly total column CO_2 observations with global coverage from a single instrument overcoming the apparent logistic problems of the ground-based network (national and private sovereignty, accessibility). As it could be seen from Figure 3.15, the combination of LEO imaging instruments and GEO capabilities by 2030 will

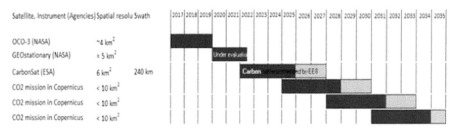

FIGURE 3.14
Future satellite missions with capabilities to resolve fossil fuel CO_2 emissions.

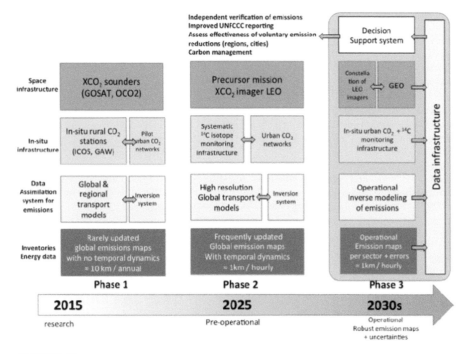

FIGURE 3.15
Three different phases in the European contribution to the construction of an operational observation system for monitoring fossil CO_2 emissions.

enhance the accuracy, timeliness, and globality of the collected data from higher altitude in space.

3.2.1.2 Early Satellite Sensing

3.2.1.2.1 U-2 Aircraft Reconnaissance Program[32]

It was after 1995 when President Clinton ordered the declassification of the airborne imagery, when the CORONA program and U-2 program were revealed. The U-2 program developed in 1954 was designed to collect information on the Soviet Union's military strengths. The following image, Figure 3.16, was taken from high resolution cameras on board of the U-2 aircraft in 1956:

3.2.1.2.2 CORONA Program

It was in August 1960 when the first successful CORONA satellite system was launched and marked the beginning of space reconnaissance and revolutionized remote sensing.

In comparison to the U-2 photograph, the image lacks detail. This is due to the ground resolution being 40 ft * 40 ft. However, the situation was rectified

FIGURE 3.16
U-2 aerial photograph of an airfield in the Soviet Union.

within a few years with the KH-4 having a 5 ft * 5 ft resolution. Table 3.3 below shows the summary of the CORONA program from its inception until 1972.

The KH-4, KH-4A, and KH-4B missions carried two panoramic cameras with a 30° separation angle as seen in Figure 3.17. This allowed for the creation of stereoscopic images, which permitted analysts to look at ground features from a three-dimensional perspective. By 1967 the KH-4s had cameras recording imagery at a 5 ft * 5 ft resolution.

3.2.1.2.3 Nimbus Satellites

The Nimbus satellites (Figure 3.18) were second-generation U.S. robotic spacecraft used for meteorological research and development. The spacecraft was designed to serve as stabilized, Earth-oriented platforms for the testing of advanced systems to sense and collect atmospheric science data. Seven Nimbus spacecraft have been launched into near-polar and sun-synchronous orbits beginning with Nimbus 1 on August 28, 1964. On board the Nimbus

TABLE 3.3

Summary of CORONA program (1959–1972)

	Camera	Launches	Recoveries	Time Period
KH-1		10	1	1959–1960
KH-2	C′ (Prime)	10	4	1960–1961
KH-3	C‴ (Triple Prime)	6	5	1961–1962
KH-4	M (Mural)	26	20	1962–1963
KH-4A	J (J-1)	52	94	1964–1969
KH-4B	J-3	17	32	1967–1972

Source: Kevin C. Ruffner, CORONA: America's First Satellite Program and USGS. "Declassified intelligence satellite photographs fact sheet 090–96" February 1998.

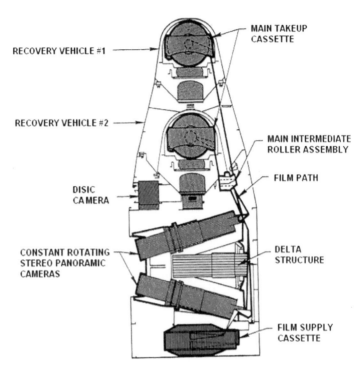

FIGURE 3.17
Stereo panoramic camera system used on KH-4A missions.

FIGURE 3.18
Nimbus satellite.

satellites are various instrumentation for imaging, sounding, and other studies in different spectral regions. The Nimbus satellites were launched aboard Thor-Agena rockets (Nimbus 1–4) and Delta rockets (Nimbus 5–7). Over a 20-year period from the launch of the first satellite, the Nimbus series of missions was the United States' primary research and development platform for satellite remote sensing of the Earth. The seven Nimbus satellites, launched over a 14-year period, shared their space-based observations of the planet for 30 years. NASA transferred the technology tested and refined by the Nimbus missions to the NOAA for its operational satellite instruments. The technology and lessons learned from the Nimbus missions are the heritage of most of the Earth-observing satellites that NASA and NOAA have launched over the past three decades.

3.2.1.2.4 *Landsat Program*

The first Landsat satellite (Figure 3.19) was launched in 1972. Since then, more satellites have been successfully launched (Figure 3.20). The Landsat satellites have repetitive, circular, sun-synchronous, near-polar orbits, providing full coverage between 81°N and 81°S. The sun-synchronous orbit means that all acquisitions over a given area occur at the same time, giving consistent shadows for the time of year. A Landsat track is 183 km wide with the repeat cycle of 16 days for Landsat 4 and later generations.

Landsats 1–3 carried the Return Beam Vidicon (RBV) camera and the multispectral scanner (MSS). The second generation of Landsat satellites,

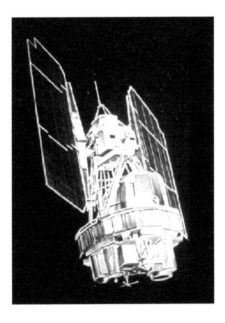

FIGURE 3.19
Landsat satellite.

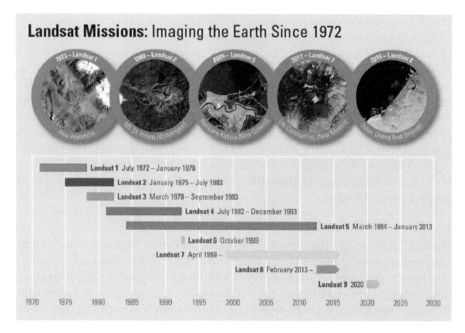

FIGURE 3.20
Timeline and history of the Landsat Missions, which started in 1972. (www.usgs.gov.)

beginning in 1982 with Landsat 4, carried a thematic mapper (TM) in addition to the MSS. Landsat 7 was equipped with an enhanced thematic mapper (ETM+). ETM+ brings a number of advantages over its predecessor, with a panchromatic band, with 15 m resolution intrinsically registered, and with the multispectral bands, while the resolution of the thermal infrared has improved to 60 m. There was also an onboard solid state recorder (SSR) with 378 GB (500 full scenes) of data capacity.

A Landsat timeline is shown in Figure 3.20 with the Landsat Data Continuity Mission (LDCM) launched in December 2012.

LDCM is the future of Landsat satellites which will continue to obtain valuable data and imagery to be used in agriculture, education, business, science, and government. The spectral sensitivities of the TM and ETM+ sensors are in sync with both the spectral response characteristics of the phenomena that the sensors are designed to monitor, as well as to the windows within which electromagnetic energy are able to penetrate the atmosphere. Table 3.4 outlines some of the phenomena that are revealed by each of the wavelengths bands, phenomena that are much less evident in panchromatic image data alone.[33]

The most recent Landsat satellite 8 (Figure 3.21) was launched in 2013, which is still operational. Next planned Landsat satellite is Landsat 9 that will be launched in 2020 as shown in Figure 3.22.

TABLE 3.4

Landsat TM/ ETM

Phenomena Revealed by Different Bands of Landsat TM/ETM Data	
Band	Phenomena Revealed
0.45–0.52 μm (visible blue)	Shorelines and water depths (these wavelengths penetrate water)
0.52–0.60 μm (visible green)	Plant types and vigor (peak vegetation reflects these wavelengths strongly)
0.63–0.69 μm (visible red)	Photosynthetic activity (plants absorb these wavelengths during photosynthesis)
0.76–0.90 μm (near infrared [IR])	Plant vigor (healthy plant tissue reflects these wavelengths strongly)
1.55–1.75 μm (mid IR)	Plant water stress, soil moisture, rock types, cloud cover versus snow
10.40–12.50 μm (thermal IR)	Relative amounts of heat, soil moisture
2.08–2.35 μm (mid IR)	Plant water stress, mineral and rock types

FIGURE 3.21
Landsat 8 satellite.

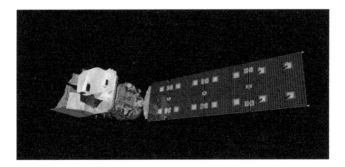

FIGURE 3.22
A rendering of the Landsat 9 spacecraft. (Image credit: Orbital ATK.)

Landsat 8 ensures the continued acquisition and availability of Landsat data, which will be consistent with current standard Landsat data products. About 400 scenes will be acquired each day. All scenes are processed to data products and are available for download within 24 hours of reception and archiving.

Landsat 8 carries two push-broom sensors: the Operational Land Imager (OLI) and Thermal Infrared Sensor (TIRS), both of which provide improved signal to noise ratio and 12-bit radiometric quantization of the data.

Table 3.5 shows different bands used in Landsats 7 and 8.

Figure 3.23 shows more than five decades of Landsat satellite developments.

3.2.1.2.5 SPOT

The Spot program was developed by the French Space Agency CNES, in cooperation with Belgium and Sweden. SPOT Image, created in 1982, is the first commercial company established to distribute geographic information derived from Earth Observation Satellites on a worldwide basis. SPOT

TABLE 3.5

Landsat 7 and 8 Bands

Satellite	Sensor(s)	Band (Wavelength, μm)	Spatial Resolution (m)
Landsat 7			
	Enhanced Thematic Mapper Plus (ETM+)	1 (0.450–0.515)	30
		2 (0.525–0.605)	30
		3 (0.630–0.690)	30
		4 (0.775–0.900)	30
		5 (1.550–1.750)	30
		6 (10.40–12.50)	60
		7 (2.080–2.350)	30
		8 (0.520–0.900)	15
Landsat 8			
	Operational Land Imager (OLI)	1 (0.435–0.451)	30
		2 (0.452–0.512)	30
		3 (0.533–0.590)	30
		4 (0.636–0.673)	30
		5 (0.851–0.879)	30
		6 (1.566–1.651)	30
		7 (2.107–2.294)	30
		8 (0.503–0.676)	15
		9 (1.363–1.384)	30
	Thermal Infrared Sensor (TIRS)	10 (10.60–11.19)	100[a]
		11 (11.50–12.51)	100[a]

Source: Table courtesy of B. Markham (July 2013). https://landsat.gsfc.nasa.gov/landsat-8/landsat-8-overview/.

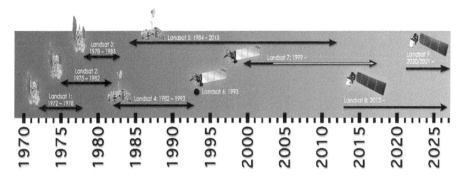

FIGURE 3.23
More than five decades of Landsat satellites.

Image's main shareholders are CNES, private industrial groups, including Matra Marconi Space which is a prime contractor for the SPOT satellites the French National Geographic Institute (IGN, 10%), and has shares held by government and private entities in Belgium, Sweden, and Italy. The SPOT payload makes up two identical HRV (high-resolution visible) imaging instruments, two tape recorders for image data, and a payload telemetry package for image transmission to ground receiving stations. Spectral bands for Spot satellites are summarized in Table 3.6 below.

In 2008, Spot Image of Toulouse, France, and partners (EADS Astrium) started an initiative to build a new commercial SPOT mission series, referred to as SPOT-6/-7, to continue sustainable wide-swath high-resolution observation services as currently provided by the SPOT-5 mission. Having full support from EADS Astrium (Spot Image's majority shareholder), manufacturing of the twin constellation SPOT-6 and -7 was officially announced in mid-2009 by Astrium Services' CEO Eric Beranger. The first launch was scheduled for 2012.

SPOT-6, launched in September 2012, is a high resolution optical Earth observation satellite delivering, 1.5 m resolution image products. In 2014, twin satellite SPOT-7 was released in to orbit with the same capabilities. SPOT-6 and

TABLE 3.6

Spot 1–5 Spectral Bands

Band	Special Range (µm)	Resolution (m) (Spot 1–3)	Resolution (m) (Spot 4)	Resolution (m) (Spot 5)
1	0.50–0.59	20	20	10
2	0.61–0.68	20	20	10
3	0.79–0.89	20	20	10
SWIR	1.58–1.75		20	20
Panchromatic	0.51–0.73	10		
Panchromatic	0.61–0.68		10	
Panchromatic	0.49–0.69			2.5/5

SPOT-7 will ensure service continuity begin with the SPOT-4 and SPOT-5 satellites, which have been operating since 1998. With a low price and extremely large catalog, this is a great option for those needing imagery for larger area sizes, especially those doing change detection and environmental analyses.

SPOT 4 payload improvements include the following:

Addition of a new band in the short-wave infrared (1.58–1.75 μm).

Onboard registration of all spectral bands. This has been achieved by replacing the panchromatic band (0.51–0.73 μm) by band B2 (0.61–0.68 μm) operating in both a 10 and 20 m resolution mode.

Improved knowledge of ground reflectance.

The two HRVIR imaging instruments are programmable for independent image acquisition, increasing significantly the total number of imaging opportunities. In particular, it is possible to change the viewing direction of one instrument without affecting the quality of the images acquired at the same time by the other instrument.

The recording capacity of each of the two onboard recorders is increased from 22 to 40 minutes. In addition, a solid-state memory of about 10 GB has been added to increase the overall reliability of onboard recording and extend the design life, while insuring a greater storage capacity.

Payload telemetry transmission to direct receiving stations is made secure to protect the commercial confidentiality of transmitted data.

Spot 5 has a new band in the 0.49–0.69 μm range.

SPOT satellites brief history:

SPOT 1 was launched February 22, 1986, with 10 m panchromatic and 20 m multispectral capability. The satellite was withdrawn from active service on December 31, 1990.

SPOT 2 was launched January 22, 1990, and is still operational.

SPOT 3 was launched September 26, 1993. An incident occurred on November 14, 1997, and after 4 years in orbit the satellite stopped functioning.

SPOT 4 was launched March 24, 1998, and includes an extra short-wave infrared band and a (low-resolution) vegetation instrument.

SPOT 5 was launched May 4, 2002, with 2.5, 5, and 10 m capability, as well as along-track stereoscopic sensors.

SPOT 6 and SPOT 7 form a constellation of Earth-imaging satellites designed to provide continuity of high-resolution, wide-swath data up to 2023. EADS Astrium took the decision to build this constellation in 2009 on the basis of a perceived government need for this kind of data. SPOT Image, a subsidiary of Astrium, is funding the satellites alone and owns the system (satellites and ground segments). Table 3.7 shows the SPOT-7 sensor specifications and Table 3.8 shows the comparison between Spot and Landsat satellites.

3.2.1.2.5.1 Spatial Resolution[34] Satellite sensors store information about objects as a grid. Digital data is collected from the area covered in the form of individual image points, so called pixels. A pixel is the smallest area unit in a digital image.

The size of the pixel is dependent on the sensor type and determines the resolution of the image. The measurement of the resolution is the edge length of a pixel. The higher the resolution and the finer the grid is, the larger is the degree of recognizable details on the Earth's surface.

The resolutions of today's satellite systems vary from a few centimeters (e.g., military usage) to kilometers. Application:

Low resolution: larger than 30 m

Medium resolution: 2–30 m

High resolution: under 2 m

TABLE 3.7

SPOT-7 Satellite Sensor Specifications

Launch date	June 30, 2014
Launch vehicle	PSLV
Launch location	Satish Dhawan Space Center (India)
Multispectral imagery (4 bands)	Blue (0.455–0.525 μm)
	Green (0.530–0.590 μm)
	Red (0.625–0.695 μm)
	Near-Infrared (0.760–0.890 μm)
Resolution (GSD)	Panchromatic – 1.5 m
	Multispectral – 6.0 m (B,G,R,NIR)
Imaging swath	60 km at Na

Source: www.satimagingcorp.com/satellite-sensors/spot-7/.

TABLE 3.8

Comparison between Spot and Landsat Satellites

	Spot	Landsat
Attitude	822 km	705 km
Inclination	98°	98.2°
Period	101.4 min	98.8 min
W track drift	2,823 km	2,752 km
Revs/day	14 + 5/26	14 + 9/16
Cycle duration	26 days	16 days
Orbits per cycle	369	233
Swath width	60 km	185 km

Different satellites are designed and launched based on their intended use and orbit. A lower resolution usually coincides with a higher repetition rate, meaning that within a short interval (METEOSAT-8, for example, every 15 minutes) the satellite investigates the same area. A "coarse" resolution is used to record large or global areas for climate-related enquiries, for example, the radiation budget of the Earth and for weather monitoring. Additional applications include Earth observation of land use and oceans, the ocean's ice cover and the surface temperatures.

Satellites with medium resolution such as LANDSAT 7 are used for the global observation of land surfaces, (Figure 3.24). Tropical rainforests and their deforestation have been observed by the LANDSAT satellites for more than 30 years.

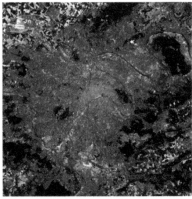

Low resolution of Western Europe - METEOSAT (1 km).   Medium resolution of Paris - Landsat 7 ETM+ (30 m).
Source: Copyright 2005 EUMETSA   Source: LANDSAT

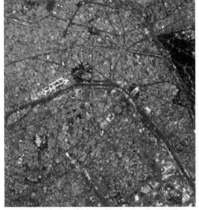

Medium resolution of a part of Paris - Landsat 7 ETM+   Medium resolution of a part of Paris - Landsat 7 ETM+
(30 m). Source: LANDSAT   (15 m pan). Source: LANDSAT

FIGURE 3.24
Low- and medium-resolution images of satellites.

FIGURE 3.25
High resolution of the Eiffel Tower – QuickBird (0.8 m).

Source: **QuickBird Satellite Image provided by GeoEye and Satellite Imaging Corporation.**

High-resolution data is mainly used for smaller areas of the Earth's surface. Only recently such data has become available commercially and privately. Satellites such as IKONOS or QuickBird send data for topographic and thematic mapping of, for example, land use, vegetation, or as planning resources for cities, large projects, etc. Information can also be "ordered" in advance, because the turning of the satellite sensors can reduce the repeat rates and can monitor the desired areas earlier.

3.2.1.3 Low-to-Medium Resolution Satellite Sensing

https://www.satimagingcorp.com/gallery/quickbird/quickbird-paris-eiffel-lg/

3.2.1.3.1 Terra Satellite

Terra (EOS AM-1) is a multi-national NASA scientific research satellite in a Sun-synchronous orbit around the Earth.[35] It is the flagship of the Earth Observing System (EOS). The name "Terra" comes from the Latin word for Earth.

The satellite was launched from Vandenberg Air Force Base on December 18, 1999, aboard an Atlas IIAS vehicle and began collecting data on February 24, 2000. It was placed into a near-polar, sun-synchronous orbit at an altitude of 705 km, with a 10:30 a.m. descending node.

Terra carries a payload of five remote sensors designed to monitor the state of Earth's environment and ongoing changes in its climate system[36]:

- ASTER (Advanced Spaceborne Thermal Emission and Reflection Radiometer)[37]
- CERES (Clouds and the Earth's Radiant Energy System)
- MISR (Multi-angle Imaging Spectro Radiometer)
- MODIS (Moderate-resolution Imaging Spectro Radiometer)[38]
- MOPITT (Measurements of Pollution in the Troposphere)[39]

Data from the satellite helps scientists better understand the spread of pollution around the globe. Studies have used instruments on Terra to examine trends in global carbon monoxide and aerosol pollution.[40] The data collected by Terra will ultimately become a new, 15-year global dataset.

3.2.1.3.2 *Aqua Satellite*

Another low-to-medium resolution remote-sensing satellite is Aqua which stands for water in Latin. Aqua is a NASA Earth Science satellite mission named for the large amount of information that the mission is collecting about the Earth's water cycle, including evaporation from the oceans, water vapor in the atmosphere, clouds, precipitation, soil moisture, sea ice, land ice, and snow cover on the land and ice.[41] Additional variables also being measured by Aqua include radiative energy fluxes, aerosols, vegetation cover on the land, phytoplankton and dissolved organic matter in the oceans, and air, land, and water temperatures. The Aqua mission is a part of the NASA-centered international EOS. Aqua was formerly named EOS PM, signifying its afternoon equatorial crossing time. Aqua was launched on May 4, 2002, and has six Earth-observing instruments on board, collecting a variety of global datasets. Aqua was originally developed for a 6-year design life, but has now far exceeded that original goal. It continues transmitting high-quality data from four of its six instruments, AIRS, AMSU, CERES, and MODIS, and reduced quality data from a fifth instrument, AMSR-E. The sixth Aqua instrument, HSB, collected approximately 9 months of high-quality data but failed in February 2003. Aqua was the first member launched of a group of satellites termed the Afternoon Constellation, or sometimes the A-Train. The second member to be launched was Aura, in July 2004, the third member was PARASOL, in December 2004, and the fourth and fifth members are CloudSat and CALIPSO, in May 2006, and the sixth member was GCOM-W1 in May 2012. In December 2013, PARASOL completed its mission and exited the A-Train. The latest addition to the A-Train was OCO-2, launched in July 2014. Now complete, the A-Train is led by OCO-2, followed by GCOM-W1, then Aqua, CALIPSO, CloudSat and, in the rear, Aura.

Aqua (EOS PM-1)

The Aqua EOS satellite is helping to better understand the causes and effects of global change. Data from its scientific instruments allows scientists to begin to piece together answers to some fundamental questions: What natural and human forces are at work? How do they interact? What can we predict? How can we prepare? What can we prevent?

Northrop Grumman built the Aqua space platform and integrated its NASA-provided science instruments. The spacecraft is based on Northrop Grumman's modular, standardized T330 bus, a design with common subsystems easily scalable to the mission-specific needs of this and future earth observing missions. Comfortable weight, power, and footprint margins readily accommodate payload replacement/refinement without affecting the basic design or the development schedule.

3.2.1.3.3 ADEOS

ADEOS (Advanced Earth Observing Satellite)[42] was a series of two Japanese Earth resources satellites. ADEOS 1, also known by its national name Midori meaning "green," was the first resources satellite to observe our planet from space in an integrated way. Developed and managed by Japan's NASDA (National Space Development Agency), it carried eight instruments supplied by NASDA, NASA, and CNES (the French space agency) to monitor world-wide environmental changes, including global warming, depletion of the ozone layer, and shrinking of tropical rainforests. Due to structural damage, the satellite went off-line after only 9 months in orbit. ADEOS 2 continued where its predecessor left off, but also studied the global circulation of energy and water. It contributed to NASA's EOS by carrying NASA's Seawinds scatterometer, a microwave radar to measure near-surface wind velocity and oceanic cloud conditions, which scientists hoped would improve their ability to forecast and model global weather. However, all communication with ADEOS 2 was lost in October 2003, probably because of heightened solar flare activity at the time. Its active lifetime had been roughly the same as that of its predecessor (Figure 3.26).

https://en.wikipedia.org/wiki/ADEOS_II#/media/File:Adeos2.jpg

Table 3.9 below shows the ADEOS 1 & 2 satellite characteristics

FIGURE 3.26
ADEOS 2. (https://aqua.nasa.gov/.)

TABLE 3.9

ADEOS 1 and 2 Satellites Characteristics

	ADEOS 1	**ADEOS 2**
Launch date	August 17, 1996	December 14, 2002
Launch vehicle	H-2	H-2A
Launch site	Tanegashima	Tanegashima
Orbit	$800 \times 800\,km \times 98.6°$	$804 \times 806\,km \times 98.7°$
Size (stowed)	$4.0 \times 4.0 \times 5.0\,m$	$4.0 \times 4.0 \times 5.0\,m$
ss at launch	About 3,500 kg	3,730 kg

3.2.1.3.4 Radarsat

Radarsat is Canada's first Earth observation satellite and the world's first operationally oriented radar satellite. With single frequency and different beam modes and positions, Radarsat is able to meet all the requirements for continuous and complete Earth monitoring and management. The Radarsat antenna operates in the C band (5.3 GHz frequency or 5.6 cm wavelength), which is able to penetrate clouds and precipitation. Radarsat transmits and receives with horizontal polarization (HH polarization). The backscattering of the incident signal is the result of changes in surface roughness and topography as well as physical properties such as moisture content and electrical properties.

The Radarsat satellite has seven SAR (Synthetic Aperture Radar) imaging options, or beam modes. Each beam mode offers a different area coverage (from roughly 50 * 50 km/scene to roughly 500 * 500 km/scene) and resolution ranging from 8m to 100 m. The Radarsat instrument also offers a range of incidence angles from 10° to 59° (always right-looking from the satellite position) on fixed positions allowing the selection of beam positions within each beam mode. Radarsat is able to capture every 24 days the same image over the same area and acquire images on a more frequent basis. For example, the ScanSAR beam mode can view a location as frequently as once a day in high latitudes, and in less than 5 days at the equator.

Radarsat can acquire SAR data from nearly any location in the world. Data collected by Radarsat is either directly transmitted to a local network station or stored on one of the two onboard tape recorders for later downlink to a Radarsat Canadian network station. Radarsat downlinks its data to the network stations at different times than other Earth observation satellites, therefore, preventing any possible conflicts with other satellites systems. The two onboard tape recorders (OBRs) are primarily used to store data over areas that do not have participating network stations. Only one tape recorder is used at a time; the second OBR acts as a backup.

SAR: The SAR is able to operate in several beam modes as shown in Figure 3.27 and Table 3.10.

- **Standard:** Seven beam modes with incidence angle ranging from 20 to 49 deg nominal, 100 km swath width and 25 m resolution.
- **Wide:** Three beam modes with varying incidence angles, 150 km swath width.
- **Fine:** Five beam modes with 50 km swath width and resolution better than 10 m.
- **Scansar:** Wide swath width (300–500 km) with a coarser resolution of 50–100 m.
- Extended mode.

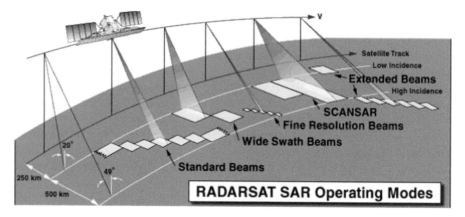

FIGURE 3.27
Different modes of SAR operation.

TABLE 3.10

Different Modes of SAR Operation

Mode	Resolution[a] (m) Range × azimuth (m)	Looks[b]	Width (km)	Incidence Angle[c] (°)
Standard	25 × 28	4	100	20–49
Wide-1	48 × 28	4	165	20–31
Wide-2	32.5 × 28	4	150	31–39
Fine resolution	11.9 × 9	1	45	37–48
ScanSAR narrow	50 × 50	2.4	305	20–40
ScanSAR wide	100 × 100	4.8	510	20–49
Extended (H)	22.19 × 28	4	75	50–60
Extended (L)	63.28 × 29	4	170	10–23

[a] Nominal; ground range resolution varies with range.
[b] Nominal; range and processor dependent.
[c] Incidence angle depends on sub-mode.

The RADARSAT Constellation is the evolution of the RADARSAT Program with the objective of ensuring data continuity, improved operational use of SAR, and improved system reliability.

The three-satellite configuration will provide daily revisits of Canada's vast territory and maritime approaches, as well as daily access to 90% of the world's surface.

The mission is currently in development, with the satellites' launch planned for 2018. SAR deserves a bit more description because of its significance in Active Remote Sensing. For airborne ground-mapping radar there has been continuous pressure and desire to achieve finer resolution. Initially, this finer resolution was achieved by the application of "brute-force" techniques.

Conventional radar systems of this type were designed to achieve range resolution by the radiation of a short pulse and azimuth resolution by the radiation of a narrow beam.

SAR is based on the generation of an effective long antenna by signal-processing means rather than by the actual use of a long physical antenna. In fact, only a single, relatively small, physical antenna is used in most cases.[43]

In considering a synthetic aperture, one makes reference to the characteristics of a long linear array of physical antennas. In that case, a number of radiating elements are constructed and placed at appropriate points along a straight line. In the use of such a physical linear array, signals are fed simultaneously to each of the elements of the array. Similarly, when the array is used as a receiver, the elements receive signals simultaneously; in both the transmitting and the receiving modes, waveguide or other transmission-line interconnections are used, and interference phenomena are exploited to get an effective radiation pattern.

The radiation pattern of a linear array is the product of two quantities if the radiating elements are identical. The radiation pattern of the array is the radiation pattern of a single element multiplied by an array factor. The array factor has significantly sharper lobes (narrower beamwidths) than the radiation patterns of the elements of the array. The half-power beamwidth B in radians, of the array factor of such an antenna is given by $B = \lambda/L$.

In this expression, L is the length of the physical array, and λ is the wavelength. In the synthetic antenna case, only a single radiating element is used in most instances. This antenna is translated to take up sequential positions along a line. At each of these positions a signal is transmitted, and the radar signals received in response to that transmission are placed in storage. It is essential that the storage be such that both amplitude and phase of received signals are preserved.

After the radiating element has traversed a distance L_{eff}, the signals in storage resemble strongly the signals that would have been received by the elements of an actual linear array. Consequently, if the signals in storage are subjected to the same operations as those used in forming a physical linear array, one can get the effect of a long antenna aperture. This idea has resulted in the use of the term *synthetic aperture* to designate this technique.

Figures 3.28 and 3.29 show images of Radarst-2 and development structure. Table 3.11 shows the RADARSAT SAR Instrument Characteristic.

RADARSAT-2 is an advanced state-of-the-art technology follow-on satellite mission of RADARSAT-1 with the objective to:

a. Continue Canada's RADARSAT program and to develop an Earth Observation satellite business through a private sector-led arrangement with the federal government.

b. Provide data continuity to RADARSAT-1 users and to offer data for new applications tailored to market needs.

FIGURE 3.28
Artist's rendition of the RADARSAT-2 satellite in orbit. (Image credit: MDA.)

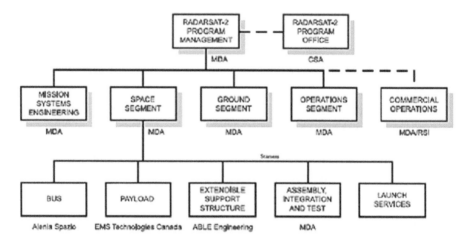

FIGURE 3.29
Overview of the RADARSAT-2 development structure. (Image credit: MDA.)

TABLE 3.11

RADARSAT SAR Instrument Characteristic

Frequency/Wavelength	5.3 GHz (C band)/5.6 cm
Polarization	Linear HH
Bandwidth	11.6, 17.3 or 30.0 MHz
Peak power	3 kW
Antenna size	15 m × 1.5 m
Incidence angle	Mode dependent
Resolution	Mode dependent

c. The key priorities of the mission respond to the challenges of:
- Monitoring the environment
- Managing natural resources
- Performing coastal surveillance

3.2.1.3.5 ERS

Launched and operated by the ESA, the ERS satellites are the first missions acquiring commercially available microwave radar data. Because microwave instruments can operate independent of the time of the day and weather conditions, it offers new opportunities for all-weather remote-sensing applications. ERS-1 operated regularly from July 25, 1991, to March 10, 2000. ERS-2 (Figure 3.30) started regular acquisition of data in May 1995, and it is still operational. Both ERS (1 and 2) satellites were launched into a sun-synchronous orbit at an inclination of 98° 52′ and an altitude between 782 and 785 km. The ERS–SAR reference system is based on a regular grid of tracks and frames. Successive tracks are counted in a continuous non-repetitive way to give a unique value for the orbit, which identifies both a track and the relative acquisition date.

Low bit-rate (LBR) data are typically global products and are provided on a yearly (or a 35-day cycle) basis.

ERS-2 ceased its operations in mid-2011, after 16 years of successful activity.

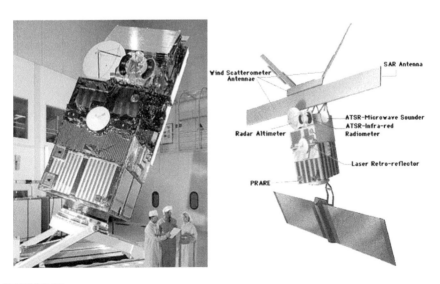

FIGURE 3.30
Photo of the ERS-2 spacecraft at ESTEC in 1994 a basic view of the deployed ERS 2 spacecraft. (Image credit: ESA, https://directory.eoportal.org/web/eoportal/satellite-missions/e/ers-2.)

3.2.1.4 *High-to-Very-High-Resolution Satellite Sensing*

The most popular high-resolution remote-sensing satellites and their characteristics are listed in the following table (Table 3.12). Resolutions are in meters.

3.2.1.4.1 *Ikonos*

Ikonos (Figure 3.31), which was launched on September 24, 1999, was the first commercial high-resolution satellite, collecting 1-m panchromatic and 4-m multispectral imagery. Although Russian sources have been providing declassified military high-resolution panchromatic data for many years, Ikonos is a wholly commercial venture, providing data to an expanding remote-sensing market. Ikonos provides full coverage of the Earth on a repetitive, circular, sun-synchronous, and near-polar orbit (PO) basis. With a maximum incidence angle of 26°, the sensor can be inclined to acquire imagery up to 700 km either side of the track, thus giving the satellite the capacity to revisit, although with different sensor angles and resolution.

TABLE 3.12

High-Resolution Remote-Sensing Satellites

Satellite	Date	Number of Bands	Max Resolution (m)
Spot 6-7	2012	4	1.50
IKONOS	1999	4	1.00
Quickbird	2001–E2014	4	0.60
Pleiades 1A-1B	2011	4	0.50
Worldview-1	2007	4	0.46
Gaofen-9	2015	1	1

FIGURE 3.31
Ikonos satellite launched in 1999.

Ikonos can acquire data over almost any area of the Earth's surface, being equipped with an onboard recorder. This recorder can hold 64 GB of data which is approximately 26 full images of both Pan and MS data. A network of ground receiving stations owned by Space Imaging's affiliates is being constructed to enable direct downlinking of data in many areas.

3.2.1.4.2 QuickBird

QuickBird is one of the world's highest-resolution commercial satellites (Figure 3.32 and Table 3.13). The QuickBird satellite collects both multispectral and panchromatic imagery concurrently. Pan-sharpened composite products (70 cm) in natural or infrared colors are also offered. DigitalGlobe's QuickBird satellite provides the largest swath width, largest onboard storage, and highest resolution of any currently available or planned commercial satellite. QuickBird is designed to efficiently and accurately image large areas with industry-leading geolocational accuracy and is capable of acquiring over 75 million km^2 of imagery data annually (over three times the size of North America).

3.2.1.4.3 Pléiades

Pléiades 1A and 1B are the newest high-resolution optical imaging satellites, operating as a constellation in the same orbit, phase 180° apart. Pléiades 1A was launched on December 17, 2011, 1B on December 2, 2012—both from Sinnamary, French Guiana. Each satellite has a maximum acquisition capacity of 1 million square km per day and has exceptional agility to maximize

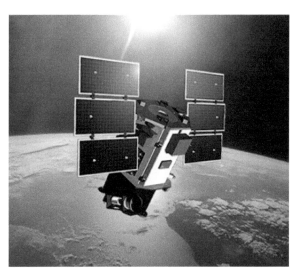

FIGURE 3.32
Quickbird satellite.

TABLE 3.13

Quickbird Satellite Design and Specifications

Launch information		Date: October 18, 2001
		Launch vehicle: Delta II
		Launch site: SLC-2W, Vandenberg Air Force Base, California
Mission life		Extended through early 2014
Spacecraft size		2,400 lbs, 3.04 m (10 ft) in length
	Altitude 482 km	Altitude 450 km
Orbit	Type: Sun-synchronous, 10:00 am descending node Period: 94.2 minutes	93.6 minutes
Sensor resolution and spectral ban width	Panchromatic: 65 cm GSD at nadir	Panchromatic: 61 cm GSD at nadir
	Black & White: 405–1053 nm	Multispectral: 2.44 m GSD at nadir
	Multispectral: 2.62 m GSD at nadir	
	Blue: 430–545 nm	
	Green: 466–620 nm	
	Red: 590–710 nm	
	Near-IR: 715–918 nm	
Dynamic range	11-bits per pixel	
Swath width	Nominal swath width: 18.0 km at nadir	Nominal swath width: 16.8 km at nadir
Attitude determination and control	Type: 3-axis Stabilized	
	Star tracker/IRU/reaction wheels, GPS	
Retargeting agility	Time to slew 200 km: 37 second	38 second
Onboard storage	128 GB capacity	
Communications	Payload data: 320 Mbps X-band	
	Housekeeping: X-band from 4, 16 and 256 Kbps, 2 Kbps S-band uplink	
Revisit frequency (at 40°N Latitude)	2.5 days at 1 m GSD or less	2.4 days at 1 m GSD or less
	5.6 days at 20° off-nadir or less	5.9 days at 20° off-nadir or less
Metric accuracy	23 m CE90, 17 m LE90 (without ground control)	
Capacity	200,000 km² per day	

acquisitions of a specific area of interest using five acquisition modes: Target, Strip Mapping, Tristereo, Corridor and Persistent Surveillance. Table 3.14 below shows the Pléiades Satellite Specifications.

3.2.1.4.4 WorldView

The next-generation commercial imaging satellite of DigitalGlobe Inc. (Longmont, CO, USA) is called WorldView-1, a successor of QuickBird-2 (launch October 18, 2001 – and fully operational as of 2008). In October 2003, DigitalGlobe was awarded a sizeable contract by NGA (National Geospatial-Intelligence Agency) of Washington DC, formerly NIMA (National Imaging

TABLE 3.14

Pléiades Satellite Specifications

	Panchromatic	**Multispectral**
Spectral range	480–830 nm	430–550 nm (blue)
		490–610 nm (green)
		510–580 nm (green)
		600–720 nm (red)
		750–950 nm (near-infrared)
Products	Color: 50 cm (merge); Bundle: 50 cm panchromatic; 2 m multispectral	
Accuracy	4.5–10 m CE 90 (exclusive of terrain displacement) with refined altitude data without Ground Control Points (GCP's)	
Swath width	20 km (single pass mosaics up to 100 × 100 km)	
Dynamic range	12-bits per pixel	
Revisit Rate	Daily	
Altitude	694 km	
Orbit	Sun-synchronous	
Stereo	Pléiades-1 is the first high resolution satellite to commercially offer Tristereo capability	

and Mapping Agency), to provide high-resolution imagery from the next-generation commercial imaging satellites.

The NGA requirements call for imagery with a spatial resolution of 0.5 m panchromatic and 2 m MS (Multispectral) data. The contract award was made within NGA's **NextView** program, designed to give the U.S. commercial imaging satellite operators the financing to build their satellites for high-resolution imaging. The WorldView mission is intended to provide imaging services to NGA as well as to the commercial customer base of DigitalGlobe.

3.2.1.4.5 GAOFEN

GAOFEN 9[44] is a remote-sensing Chinese satellite. The spacecraft was tracked in an orbit of 617 * 664 km inclined 98°, in family with previous Gaofen Earth observation satellites as well as the military Yaogan satellite series which is suspected to be the origin of the two Gaofen satellites launched in recent months. According to official reports, the Gaofen-9 satellite is capable of delivering imagery at a resolution of better than one meter. Imagery from the satellite is to be used in urban planning, road network design, land surveying, and other purposes. The Gaofen satellites are part of the High-Resolution Earth Observation System, CHEOS that had originally included plans for seven satellites to create a constellation of spacecraft readily providing data products from a variety of sensors to establish an operational Earth observation capability similar to that of the American Landsat and European Copernicus programs. CHEOS satellites will be equipped with a variety of sensors from high resolution, multispectral optical imagers to radar payloads, infrared imagers and microwave payloads as well as specific sensors. Satellites belonging to CHEOS will be operated in different orbits including LEO and Geostationary Orbit to create a range of data products at different qualities and revisit times.

NORAD ID: 40894; **Int'l Code**: 2015-047A; **Perigee**: 623.6 km; **Apogee**: 671.8 km; **Inclination**: 98.0°; **Period**: 97.5 minutes; **Semi major axis**: 7,018 km; **RCS**: Unknown; **Launch date**: September 14, 2015; **Source**: People's Republic of China (PRC); **Launch site**: Jiuquan Satellite Launch Center, China (JSC).

3.2.1.5 Indian Remote-Sensing Satellite

The Indian Remote Sensing (IRS-1C) satellite was successfully launched into PO on December 28, 1995, by a Russian launch vehicle. Its payload was activated in the first week of January 1996. This satellite was followed by a similar one that was successfully launched into PO on September 29, 1997, by a polar satellite launch vehicle (PSLV). Its payload was activated in mid-October 1997.

The primary objective of IRS satellites is to provide systematic and repetitive acquisition of data of the Earth's surface under nearly constant illumination conditions. IRS-1C operates in a circular, sun-synchronous, near-PO with an inclination of 98.69°, at an altitude of 817 km in the descending node. The satellite takes 101.35 minutes to complete one revolution around the Earth and completes about 14 orbits per day. The entire Earth is covered by 341 orbits during a 24-day cycle.

As of February 2016 the status of India's remote-sensing satellites could be summarized as follows:

- Three satellites in Geostationary orbit (INSAT 3D, Kalpana, and INSAT 3A)
- Nine satellites in Sun-synchronous orbit (RESOURCESAT-2; CARTOSAT-1, 2, 2A and 2B; RISAT-1 and 2; OCEANSAT 2; SARAL)
- Equatorial orbit (MEGHA TROPIQUES)
- Both Optical and Microwave Sensors providing wide range of spatial, spectral, radiometric, and temporal resolutions.

Cartosat-2 Series Satellite is the primary satellite carried by PSLV-C40. This remote-sensing satellite is similar in configuration to earlier satellites in the series and is intended to augment data services to the users.

The imagery sent by satellite will be useful for cartographic applications, urban and rural applications, coastal land use and regulation, utility management like road network monitoring, water distribution, creation of land use maps, change detection to bring out geographical and manmade features and various other Land Information System (LIS) as well as Geographical Information System (GIS) applications.

PSLV-C40/Cartosat-2 Series Satellite Mission was launched on January 12, 2018 at 09:29 hours (IST) from SDSC SHAR, Sriharikota.

Launch Mass: 710 kg; Launch Vehicle: PSLV-C40/Cartosat-2 Series Satellite Mission; Type of Satellite: Earth Observation; Manufacturer: ISRO; Application: Earth Observation; Orbit Type: SSPO.[45]

3.2.1.6 National Oceanic and Atmospheric Administration

NOAA operates two types of satellite systems for the United States: geostationary satellites and polar-orbiting satellites (www.noaa.gov/satellites. html). Geostationary satellites constantly monitor the Western Hemisphere from around 22,240 miles above the Earth, and polar-orbiting satellites circle the Earth and provide global information from 540 miles above the Earth. Satellites enable us to provide consistent, long-term observations, 24 hours a day, 7 days a week. They track fast-breaking storms across "Tornado Alley" as well as tropical storms in the Atlantic and Pacific oceans. Data from satellites are used to measure the temperature of the ocean, which is a key indicator of climate change. Satellite information is used to monitor coral reefs, harmful algal blooms, fires, and volcanic ash. Monitoring the Earth from space helps us understand how the Earth works and affects much of our daily lives. NOAA's satellites provide other services beyond just imaging the Earth. Monitoring conditions in space and solar flares from the sun help us understand how conditions in space affect the Earth. Satellites also relay position information from emergency beacons to help save lives when people are in distress on boats, airplanes, or in remote areas. Scientists also use a data collection system on the satellites to relay data from transmitters on the ground to researchers in the field.

NOAA has also been operating its Advanced TIROS-N (ATN) series of satellites since October 1978. It carries the TIROS Operational Vertical Sounder (TOVS), a low data-rate atmospheric sounding package, and the advanced very-high resolution radiometer (AVHRR). The satellites were designed primarily for meteorological applications, but the AVHRR instrument, in particular, has proven very valuable in a variety of environment-linked Earth observation applications; therefore, its services and applications have expanded ever since. The TIROS satellites have a circular, polar, sun-synchronous orbit, with an altitude of 840 km and an inclination of 98.7° to 98.9°. NOAA maintains at least two operational satellites in complementary orbits, with the even-numbered satellites crossing the equator at local solar times of approximately 07:30 and 19:30, and the odd-numbered satellites at 02:30 and 14:30.

3.2.1.7 Russian Remote-Sensing Satellites

The Commonwealth of Independent States (CIS)—the former Soviet Union's space activities show great and expanding interest in Earth observation, not only for military purposes, but also for assessing resources on a regional and global scale.

METEOR SERIES (2-21, 3-5, 3M)—Meteor was the first civil applications satellite deployed by the former USSR. It is comparable, in many ways, to the U.S. NOAA series. Numbers of meteor spacecraft were launched and then replaced. The new meteor satellites are designed to accommodate a variety of payload packages. For example, Meteor 3 carried NASA's Total Ozone Mapping Spectrometer (TOMS).

OKEAN—A series of all-weather oceanographic satellites with real aperture side-looking radars. These satellites provide all-weather monitoring of ice conditions, wind-induced seaway, storms and cyclones, flood regions, and ocean surface phenomena.

RESURS—The Resurs series spacecraft are roughly comparable to the U.S. Landsat system. They carry a multispectral instrument package, operating in the visible to thermal infrared regions, and have been known for their ability to detect industrial pollution.

ELECKTRO (GOMS)—The Geostationary Operational Meteorological Satellite carries a sensor suite similar to the NOAA GOES satellite series. In real time, it obtains visible and infrared images of the Earth surface and cloud cover, and provides continuous observation of the dynamics of varying atmospheric processes on Earth.

Although no longer in service, very important sources of remote-sensing data were the ALMAZ satellite, and Mir space station.

Currently, a group of eight Russian remote-sensing satellites is in orbit around the planet. The number of operating Russian remote-sensing satellites orbiting the Earth will reach 15 by 2020, according Russian President. That will allow imaging the Russian territory as well as the whole Earth.

3.2.1.8 Issues and Remarks

The development of high-resolution satellites and the authorization to own and operate them with virtually no restrictions has stimulated an unprecedented competition by large aerospace companies to create a new geospatial infrastructure aimed at markets never before considered. Numerous Earth observing satellite systems having spatial resolutions of 0.8–5 m in the panchromatic band and of 3.3–20 m in three to ten or more multispectral bands are stimulating great interest, curiosity, and conjecture by the governmental, scientific, and public communities. The outstanding quality of imagery promised by these companies is due to technological advancements in computers, sensors, processing, and communications which can be mainly attributed to advanced digital imaging technology developed for military

Sensor	Resolution (m)	Spectral range (μm)	Rev. int. (days)
Landsat	15	0.45 to 2.35	16
SPOT	2.5	0.50 to 1.75	5
IRS	5	0.50 to 1.70	5
Ikonos	1	0.45 to 0.85	3
Quickbird	0.61	0.45 to 0.90	3
FORMOSAT	2	0.45 to 0.90	1
CARTOSAT	2.5	N/A	5
Worldview	0.46	0.40 to 1.04	1.1
ALOS	2.5	0.42 to 0.89	2
Geoeye	0.41	0.45 to 0.90	3
Airborne	1 to 25	0.42 to 14.00	N/A

FIGURE 3.33
Summary of the sensor properties of remote-sensing satellites.

and intelligence organizations during the Cold War era. Figure 3.33 illustrates the summary of the sensor properties of remote-sensing satellites.

3.2.1.9 Resolution Restrictions and Legislative Actions

The principles relating to remote sensing of the Earth from space adopted by the UN on December 3, 1986, have proven to be insightful for guiding the impending commercialization of remote sensing. Its comprehensive coverage has minimized the extent of legal issues regarding Earth observation, including open skies for high-resolution systems. These principles were adopted through the efforts of the UN Committee on the Peaceful Uses of Outer Space (COPUOS) as part of the progression of formulating international rules to enhance opportunities for international cooperation in space.
   Remote-sensing activities of a state shall:

1. Be for the benefit and in the interest of all nations.
2. Be conducted in accord with international law.
3. Be conducted with respect for sovereignties and rights of sensed states.
4. Promote international cooperation in an equitable manner.
5. Encourage establishment of regional agreements for data collection and processing where feasible.
6. Make available technical assistance to interested states on mutually agreed terms. The UN shall promote international cooperation, including technical assistance and coordination.

7. Inform UN Secretary General of and about its space programs.
8. Promote protection of Earth environment and inform states affected.
9. Promote protection from natural disasters and inform states likely to be affected.
10. Make data accessible to sensed states on nondiscriminatory basis and on reasonable cost terms.
11. Consult with sensed states for mutual opportunities.
12. Be responsible and adhere to international law.
13. Resolve disputes from application of these principles through established procedures.

These principles, however, were formulated and established when governments were the sole operators of remote-sensing satellites and most nations had 10–50 m resolution restrictions for civil Earth observation satellites. Today, with the absence of resolution limits in the U.S. legislation for commercial systems, "the genie is out of the bottle!" What was once the privileged domain of a few defense and intelligence agencies is now available for civil and commercial applications.

It is the intention of most new commercial systems to authorize consumers of their imagery and imagery products for single-use, single-purpose purchases. However, as with any commodity, it is very difficult to identify and control the use of a product. The UN Principles place the international legal responsibility on the government of the image providers for the control of their activities. Undoubtedly, new national and local legislation will be needed to regulate unlawful use of space-derived imagery.

3.2.1.10 Market Development

Although the traditional markets for Earth observation imagery have been weather prediction and monitoring, as well as some surveying and mapping, the future market is for spatially attributed temporal information. A market niche in the transportation sector for real-time navigation has already been spawned for use of dynamic map displays generated from satellite imagery data combined with GPS and inertial referencing systems. Another very promising market sector is precision farming, which uses timely repetitive imagery. The many other promising niche and spin-off markets envisioned for spatial information systems include, but are surely not limited to, the following:

Disaster monitoring and assessment services
Emergency services
Tracking hazardous activities
Fire and hazards detection

Disease detection (agricultural)

Disease monitoring (agricultural and human)

Real estate appraisal, taxation, and permitting

City and urban planning

Facilities placement and monitoring

Peacekeeping and treaty monitoring

Law enforcement

News gathering services

Environmental protection

Resource monitoring and assessment (natural and renewable)

Archaeological and architectural site monitoring and preservation

Trends analysis and prediction services

Navigation safety

Reconnaissance, detection, and surveillance

Demography

3.2.2 Meteorological Satellites

From the early 1960s, the weather, or meteorological satellite programs have been of interest to several organizations in United States, Europe, Russia, and the rest of the world. In the United States, NASA, NOAA, and the Department of Defense (DOD) have been developing and operating weather satellites. In Europe, the ESA and European Organization for Exploration of Meteorological Satellites (EUMETSAT) operate the meteorological satellite systems. TIROS[46] (Figure 3.34) was the world's first weather satellite launched by NASA on April 1, 1960, as an experimental satellite to study the Earth. The improved TIROS Operational System (ITOS) began operating in 1970, later named NOAA.

The main objectives of these LEO (833–870 km) satellites were to provide improved infrared and visual observations of Earth's cloud cover for use in analyzing weather and forecasting. NASA's Nimbus satellites launched from 1964 through 1978 practically became weather satellites, providing data to the Environment Science Services Administration (former name for the National Weather Services), although it was originally designed for testing new sensing instruments and data-gathering techniques. Instruments on the Nimbus satellites included microwave radiometer, atmosphere sounders, ozone mappers (TOMS, on Nimbus 7), the Coastal Zone Color Scanner, and infrared radiometers. These instruments on board Nimbus satellites provided a wide range of crucial global data on sea-ice coverage, atmosphere temperature, atmosphere ozone distribution, and the amount of radiation in the Earth's atmosphere and sea-surface

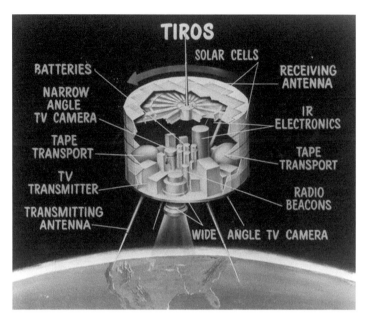

FIGURE 3.34
TIROS-1 construction. (www.pinterest.com/pin/456622849700758499/.[46])

temperature. In addition to weather satellites operating in the LEO, such as those mentioned above, there are satellites that operate in geostationary orbit (22,400 miles, 35,700 km) above the Earth's equator. Geosynchronous weather satellites provide continuous day and nighttime monitoring of almost the entire hemisphere. The first geosynchronous meteorological satellite (SMS-1) was launched by NASA on May 17, 1974. GOES-1[47] followed SMS-1 by its launch on October 16, 1975. Figure 3.35 shows the GOES coverage area of the GOES two-satellite constellation that views nearly 60% of the Earth's surface.

Once the satellite is placed in its orbit, it experiences various perturbing torques including gravitational forces from other bodies like solar and lunar attraction, magnetic field interaction, solar radiation pressure, and so forth. Because of such effects, the satellite orbit tends to drift which causes its orientation to change. To remain in the allowable "box" of the orbital allocation, the satellite's position needs to be controlled both in the east–west as well as the north–south directions. The east–west position needs to be maintained to prevent any possible radio frequency (RF) interference from the adjacent satellites. For the satellite to have the proper inclination, the north–south orientation also has to be maintained. The satellite position and its orientation are maintained through the attitude and orbit control system which also

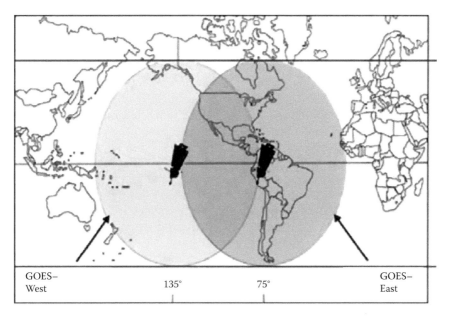

FIGURE 3.35
GOES coverage area.

keeps the antenna correctly pointed in the desired direction. When needed, the orbital control is maintained by firing thrusters in the desired direction or by releasing a jet of gas.

The techniques used to achieve the satellite stabilization are spin (or body) stabilization and three axis stabilization. Figure 3.36[48] shows the spin

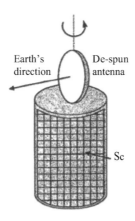

FIGURE 3.36
Spin stabilization.

stabilization, and Figure 3.37[49] shows the three axis stabilization methods. The former is mostly used in early generations of satellites and the latter in more recent satellites.

GOES-8 was the first weather satellite using the three-axis stabilization (Figure 3.38)[50] that resulted in noticeable improvement in gathering weather imagery and atmosphere data. The most recent GOES satellites use GPS for search and rescue operations. In 1969, the French national space agency, Centre National d-Etudes Spatiales (CNES) proposed the European meteorological satellite system, Meteosat (Figure 3.39).[51] On November 23, 1977, Meteosat-1 was launched which was followed by the launch of Meteosat-2 in 1981. EUMETSAT, Europe's meteorological satellite organization, which is an intergovernmental organization, was created through an international convention agreed to by 17 European member states in 1986. EUMETSAT and ESA have been cooperating on the production of a significantly improved system.

Figure 3.40 shows the Meteosat first-generation and second-generation satellite characteristics that demonstrate the improvements made. The year 1997 marked the launch of the Meteosat-7, the last of the first-generation weather satellites developed by ESA. The operation of the Meteosat satellites was formally handed over from ESA to EUMETSAT in 1995. The main services initially provided by Meteosat's first-generation satellite have

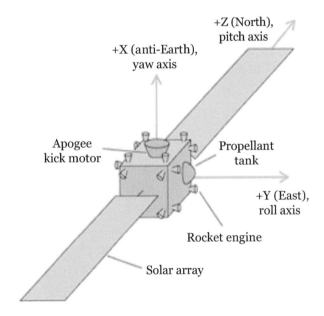

FIGURE 3.37
Three axis satellite stabilization.

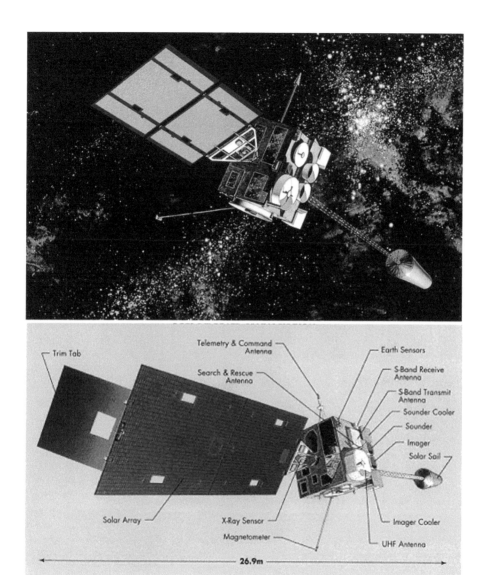

FIGURE 3.38
The components of a deployed GOES satellite are labeled.

been replaced by the improved Meteosat second-generation (MSG), with two currently in orbit, Meteosat-8 and Meteosat-9, and two more satellites planned for 2021. The Meteosat program is set to continue with MGS-3, which will be launched in mid-2012 on an Ariane-5, followed by MSG-4 in 2014.

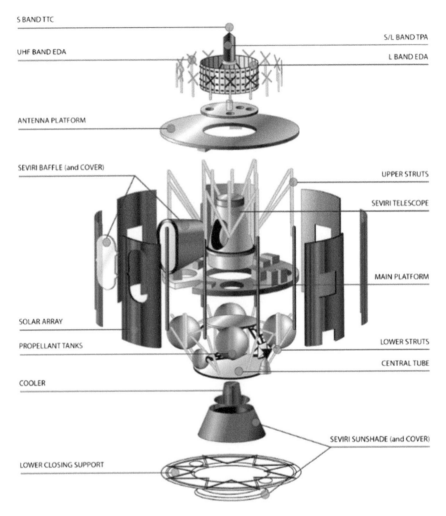

S BAND TTC

S/L BAND TPA

UHF BAND EDA

L BAND EDA

ANTENNA PLATFORM

SEVIRI BAFFLE (and COVER)

UPPER STRUTS

SEVIRI TELESCOPE

MAIN PLATFORM

SOLAR ARRAY

PROPELLANT TANKS

LOWER STRUTS

CENTRAL TUBE

COOLER

SEVIRI SUNSHADE (and COVER)

LOWER CLOSING SUPPORT

FIGURE 3.39
Blowup of the MSG structure showing the main components of the S/C and its payload.
(Image credit: EUMETSAT.)

The building blocks for the MSG ground segment are shown in following figure:

The EUMETSAT central processing facility in Darmstadt produces a wide range of meteorological products derived from satellite data. Some of these products are as follows:

Atmospheric wind vectors at various altitudes

Clouds analysis providing identifications of cloud layers with coverage, height, and type of weather forecasting

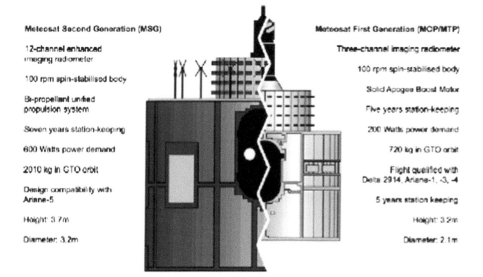

FIGURE 3.40
Comparison of first and second generation Meteosat spacecraft. (Image credit: EUMETSAT.)

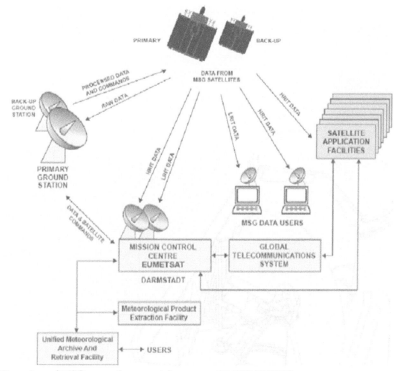

Overview of MSG ground segment (image credit: EUMETSAT)

Tropospheric humidity at medium and upper levels for weather forecasting

High-Resolution Precipitation Index (HRPI)

Cloud Top Height images for aviation meteorology

Clear Sky Radiances, for numerical weather prediction models

Sensors used onboard of meteorological satellites could be classified into two types: active and passive. Active sensors provide their own source of illumination and the information about the target is deduced from the reflected radiations from the targets.

The second type of sensors used on satellites are passive sensors in which the source of illumination is not provided by the sensor, and these sensors measure the emitted radiation and reflected radiation from the targets of interest.

Figure 3.41 shows sensors used in meteorological satellites and their modes of operation. NOAA satellites referred to earlier carry the following meteorological payloads:

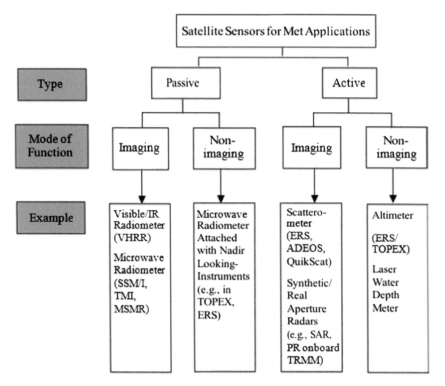

FIGURE 3.41

Sensors used in meteorological satellites and their modes of operation. (Kishtawal, C.M., Meteorological Satellites, Atmospheric Science Division, Meteorology and Oceanographic Group, Space Application Center (ISRO), Ahmedabad, India.[52])

AVHRR

TOVS

Earth Radiations Budget (ERB)

3.2.3 Global Positioning Satellites

The use of satellites for navigation traces back to early 1959 when the Navy navigation satellite system was developed and introduced into service in 1964. GPS has found many applications in military and civil sectors during the last five decades. Military applications included general navigation, weapon targeting, and mine countermeasures. The general navigation of a wide range of vehicles, from tanks to ships to jet fighters, provides accurate position information that is vital for efficient command, control, communication, and intelligence (C^3I) aspects of military operations. GPS application in weapon targeting helps to accurately find the positions of artillery pieces for precision targeting.

Civil applications include general navigation for a wide range of civil vehicles, from luxury cars to super tankers. The offshore oil industry also utilizes the accurate position determination of GPS for seismic surveying, rig positioning, and pipe laying. In these types of applications, in order to provide real-time accurate information, the GPS is used in differential mode.

GPS was developed by the U.S. Department of Defense and reached its full operating capability on July 17, 1995. GPS operates by using a constellation of medium earth orbiting satellites and a control station in Colorado Springs in addition to five monitoring stations and their antennas, all spread across the Earth. When GPS became available for civilian use, the DOD introduced an error called selective availability (SA), which limited GPS to approximately 100 m of accuracy for national security reasons. This selective availability was eliminated in May 2000, providing a current accuracy of approximately 10 m.

To provide worldwide coverage, 24 satellites are needed (Figure 3.42). These satellites that operate from the medium earth orbits (20,200 km) provide accurate position information (latitude and elevation) to users on land, in the air, or at sea, globally with an orbital period of 12 hours.

The GPS operates by using triangulations with a minimum of four satellites, although three satellites are adequate if altitude position is known, as in the case of ships at sea. The received signal level is usually below the noise power at the receiver, but spread spectrum techniques are used to improve the received signal-to-noise ratio. GPS operates at two frequency bands: L1, at 1,575.42 MHz; and L2, at 1,227.60 MHz, transmitting spread spectrums signals with binary phase shift keying modulation.

The L1 frequency is used to transmit the positions of the satellites (ephemeris) as well as timing codes, which are available to any commercial or public

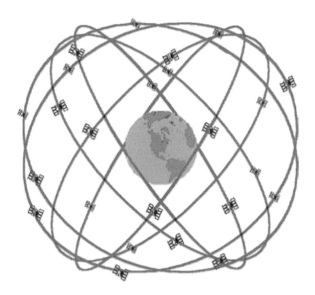

FIGURE 3.42
GPS satellite system. (Peter H. Dana, Global Positioning System Overview, www.colorado.edu/geography/gcraft/notes/gps/gps_f.html.[53])

user. This mode of operation is referred to as the Course/Acquisition (C/A) code. The L2 frequency is however reserved for military use and uses an encrypted timing code referred to as the Protested (P) code. The P code offers much higher accuracy than the C/A code. The GPS system uses one-way transmission from satellites to users, so that the user does not require a transmitter, only a GPS receiver. Because the free space value for the speed of light for radio waves traveling through the atmosphere and the troposphere are slightly different than that through vacuum, it can introduce timing errors in certain applications. In addition to this error, the satellite clocks also have their own timing error although they are highly accurate. Adding to this error, another error is caused as a result of the satellite geometry, known as dilution of precision (DOP). The dilution of position errors combined with the above timing errors set a limit on the accuracy of location determination. In cases when very high accuracy is required, differential GPS (DGPS) can be used. In this mode, two receivers are used, one of which is placed at an accurately known location.

The GPS system mainly consists of three key segments: the space segment, the control segment, and the user segment. Figure 3.43 shows the GPS configuration.

The space segment as was described earlier consists of 24 satellites in a sun-synchronous orbit with a period of about 12 hours (11 hours 58 minutes) at an altitude of 20,200 km. The satellites are placed into six planes, each having four satellites and separated by 60°. The control segment is made of the master control station and a number of monitor station and uplink stations. All the necessary functions needed to manage the operation of the GPS systems are carried

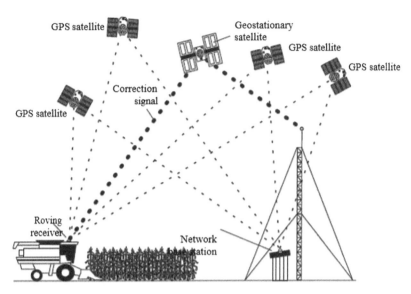

FIGURE 3.43
Conceptual rendition of GPS satellite differential correction communication for agricultural equipment positioning. (Peter H. Dana, Global Positioning System Overview.)

out by the control segment. The user segment consists of a wide spectrum of user receivers, each tailored toward a different application as was described earlier. Some of the applications of GPS are listed below; however, new applications emerge continuously as the potential of GPS is better understood.

Agriculture—By using the combination for GPS and GIS (geographic information system) data, more precise information is obtained on site-specific farming which is used for farm planning, variable rate applications, field mapping, soil sampling, tractor guidance, crop scouting, and yield mapping. When remote-sensing data is added to the GPS–GIS data, even more accurate results are obtained about the target under study. Figure 3.43 also shows the GPS application for agricultural equipment positioning.

Aviation—GPS is used to increase the safety and efficiency of flights. As new and more efficient routes are made possible by using GPS, more savings in time and money are being realized. In the aviation community, the good news is that the GPS is being constantly improved in order to reduce errors due to disturbances in the ionosphere that improves accuracy, availability, and integrity of the system.

Environment—GPS technology helps to understand and forecast changes in the environment and by integrating its measurements into meteorological data, the atmospheric water content can be determined which helps to improve the accuracy of weather forecasts.

GPS information is also used in tracking the movement and spread of oil spills, mapping the perimeter of forest fires, helping scientists to anticipate earthquakes, vegetation mapping, and tropical rain forest ecology.

Marine—Providing an accurate method for navigating, measuring speed, and determining location makes GPS a very attractive technology to the marine environment. Knowing a ship's position while in open sea and also in congested harbors and waterways is very important in marine navigation. Mariners and oceanographers are also using GPS data for variety of applications ranging from underwater surveying to commercial fishing. Differential GPS, as described earlier provides even higher precision and safety in marine operations.

Public safety and disaster relief—GPS has been instrumental in the global disaster relief efforts such as the 2004 Indian Ocean region tsunami and 2005 Hurricane Katrina. GPS has also been used to manage forest fires, help scientists to anticipate earthquakes tracking storms, predict floods, and in short, GPS helps to save lives in a wide variety of disastrous situations.

Rail, roads, and highways—In the rail system, GPS provides information on the location of trains which helps to prevent collisions and maintain smooth flow of traffic on roads and highways. GPS provides in-vehicle navigations systems and capability of tracking bus, rails, and other services in order to improve on-time performance. GPS is also used to survey roads and highway networks provide information about the traffic, traffic violations, or crashes. In intelligent transportation systems (ITSs), to be implemented in the future, GPS will play a major role.

3.2.4 Scientific Research Satellites

Scientific satellites have the mission of capturing data and providing the scientists with vital information about Earth and our universe. Satellites of this category either look at Earth and its atmosphere or look away from Earth toward outer space. Earth-looking scientific satellites provide information about the Earth's history, its present condition, and the future developments. NASA's first satellite in this series was the Upper Atmospheric Research Satellite (UARS), launched in 1991 to investigate the mechanisms controlling Earth's upper atmosphere. This satellite provided the chlorine monoxide map and its direct link with ozone presence and depletion. The second class of scientific satellites is the satellites that look away from Earth; they investigate the sun, stars, planets, and other phenomena in our universe. Viking spacecraft launched in 1976 was able to take close-up pictures from the surface of Mars. Other such satellites include the International Ultra-violet

Explorer (IUE) that studied the intense ultraviolet output from an exploding star, supernova that revealed the rapid cooling off of the explosion. Scientific satellites have made it possible to learn more about how the universe was created. In the following, more details are given about some of these satellites.

1960s

Project Mercury—The aim of this project was to learn if humans could survive the rigors of spaceflight. Project Mercury began in 1958 and ended in 1963 and was the U.S. first man-in-space program (Figure 3.44[54]).

Project Gemini—This manned space program was announced in January 1962 and its ten flights that took place during 1965 and 1966 provided NASA scientists and engineers with weightlessness data; procedures for reentry and splash over; and rendezvousing and docking in space.

Project Apollo—This project became NASA's priority in 1961 aiming at showing U.S. superiority over the Soviet Union during the Cold War. This project lasted 11 years and cost $25.4 billion. Apollo 8 was the first satellite to orbit the moon. Apollo 11s (Figure 3.45[55]) primary objective was to complete a nation goal set by President John F. Kennedy on May 25, 1961, to perform a crewed lunar landing and return to Earth. The other objective of this flight was the scientific exploration by the Lunar Module (LM) crew, deployment of a television camera to transmit signals to Earth. On March 16, 1962, the Soviet Union launched COSMOS-1 and declared it a research satellite with numerous scientific research objectives. This satellite, which was also called Sputnik 11, employed radio methods to study the structure of the atmosphere.

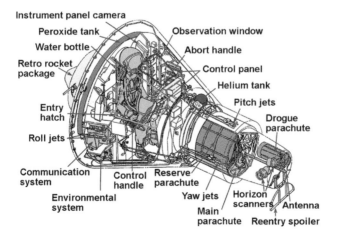

FIGURE 3.44
Mercury spacecraft systems.

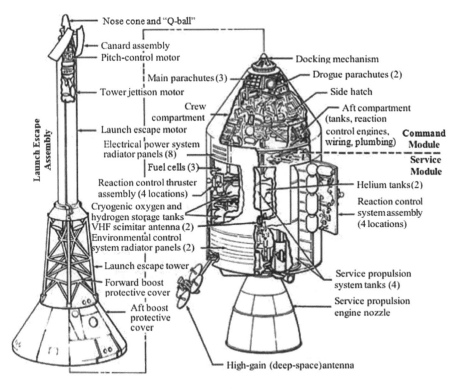

FIGURE 3.45
Apollo command and service modules and launch escape system.

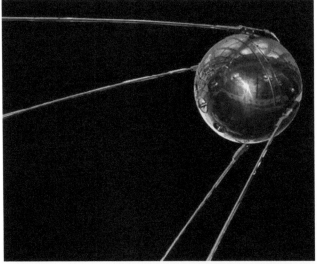

October 4, 1957, Sputnik 1 (Спутник-1), the first man-made, Earth-orbiting satellite, launched skyward. (Soviet Web Space, Sputnik: A History of the Beginning of the Space Age, 56.

1970s

Japan's first scientific satellite, Ohsumi, was launched in 1970. Its main objectives were to study satellite launch technologies by M (Mv) rocket and perform an engineering experiment on the satellite.

NASA's 1978 launch of Nimlus-7 marked a new beginning in space-based atmosphere research. The satellite contained eight Earth-viewing sensors including the first ERB experiment and TOMS instruments.

In 1972, NASA launched the Earth Resource Technology Satellite (ERTS-1) into orbit. ERTS-1 which was later named Landsat-1, used advanced instruments to view the Earth's surface in several infra-red wavelengths. These sensors provided data to help scientists to assess vegetation growth and monitor the spread of cities and other changes on the surface of the Earth.

In 1978, NASA also launched SeaSat to observe the ocean using SAR. SAR is a ground mapping method that uses computer processing to provide high-resolution images by combining radar images made many miles apart.

1980s

On November 28, 1983-STS-9 carried ESA Spacelab (which is an orbital laboratory) payload that carried 71 experiments. The investigations that were carried out were in the areas of astronomy, physics, atmospheric physics, Earth observations, life sciences, materials sciences, space plasma physics, and technology.

In 1988, the first operational Russian spacecraft (Okean-O) was launched with the following mission:

To study the world ocean as a global damper and regulator of heat and moisture content of the atmosphere

To ensure the safety of navigation and control of the ice situation in the Arctic and Antarctic

To study the dynamics of sea currents, and the process of self-purification of sea water and cleansing of river effluents

To control the intensity of pollution of the oceans with oil and oil product discharges

1990s

In September 1991, NASA launched the UARS from the space shuttle. UARS was designed to study the upper regions of the atmosphere where sounding balloons and airplanes cannot reach.

In 1992, an Ariane 42P rocket launched a spacecraft named Topex/Poseidon which was a joint project between the French Space

Agency and NASA. The radar altimeter on the satellite measured ocean topography and allowed scientists to map ocean circulation which helps to understand global weather and climate change.

In 1997, SeaStar was launched to study biological organisms in the oceans such as algae and phytoplankton (microscopic marine plants).

In January 27, 1999, the first scientist satellite of Taiwan, ROCSAT-1 was launched in order to develop the emerging technology of space research and related applications. ROCSAT carried a k_a-band experimental communications payload, an ocean color imager experiment to study plankton distribution for fisheries management, and an instrument to measure thermal plasma in the equatorial ionosphere.

2000s

In 2002, the ESA launched a large environmental monitoring satellite named Envisat, as was referred to earlier, aboard an Ariane 5 rocket. Envisat, the successor to ERS-1 and 2, is designed to take simultaneous readings of various atmospheric and terrestrial features and help in understanding the global change. To meet the mission requirement, the following instruments were selected:

Michelson Interferometer for Passive Atmospheric Sounding (MIPAS)

Global Ozone Monitoring by Occultation of Stars (GOMOS)

Scanning Imaging Absorption Spectrometer for Atmospheric Cartography (SCIAMACHY)

Medium Resolution Imaging Spectrometer (MERIS)

Advanced Along Tack Scanning Radiometer (AATSR)

Advanced Synthetic Aperture Radar (ASAR)

Radar Altimeter 2 (RA-2)

Microwave Radiometer (MWV)

Doppler Orbitography and Radio positioning Integrated by Satellite (DORIS).

Retro Reflector Array (RRA)

In October 19, 2006, the MetOp-A was launched. This satellite was equipped with 11 scientific instruments to scan the Earth's atmosphere, land, and oceans as part of a program to improve weather forecasting. MetOp is operated by the European Organization for the Exploration of Meteorological Satellites and is part of the EUMETSAT Polar System. The following instruments were onboard the MetOp-A satellite:

IASI (Infrared Atmospheric Sounding Interferometer)

MHS (Microwave Humidity Sounder)

GRAS (Global navigation satellite system Receiver for Atmospheric Sounding)

ASCAT (Advanced Scatterometer)

GOME-2 (Global Ozone Monitoring Experiment-2)

AMSU-A1/AMSU-A2 (Advanced Microwave Sounding Units)

HIRS/4 (High Resolution Infrared Radiation Sounder)

AVHRR/3

A-DCS (Advanced Data Collection System)

SEM-2 (Space Environment Monitor)

SARP-3 (Search and Rescue Processor)

SARR (Search and Rescue Repeater)

On February 16, 2007, five THEMIS satellites were launched by a DELTA II rocket from Cape Canaveral, Florida. The main objective of THEMIS is to determine the place of release and the nature of the macroscopic process responsible for the substorms, and to study their relations and couplings. Another objective was to study the radiation belts, the magnetopause, and the border layers. Each of the satellites is equipped with five instruments:

FGM (FluxGate Magnetometer)

EFI (Electric Field Instrument)

ESA (electrostatic analyzer)

SCM (search coil magnetometer)

SST (solid-state telescopes)

2010s

On November 20, 2010, RAX (Radio Aurora Explorer) also known as USA 218, was launched. This is a three-unit CubeSat that will conduct studies in the ionosphere. RAX is a 60-pound nanosatellite that resulted from a joint effort of the University of Michigan and SRI International.

In November 2013, **Lem** which is the first Polish scientific artificial satellite was launched as part of the Bright-star Target Explorer (BRITE) program. The spacecraft was launched aboard a Dnepr rocket. Named after the Polish science fiction writer Stanisław Lem, it is an optical astronomy spacecraft built by the Space Research Centre of the Polish Academy of Sciences and operated by Centrum Astronomiczne im. Mikołaja Kopernika PAN; one of two Polish contributions to the BRITE constellation along with the Heweliusz satellite.

In August 2014, **Heweliusz** which is the second Polish scientific satellite was launched as part of the BRITE program. The spacecraft was launched aboard a Chang Zheng 4B rocket. Heweliusz is an optical astronomy spacecraft built by the Space Research Centre of the Polish Academy of Sciences and operated by Centrum Astronomiczne im. Mikołaja Kopernika PAN; it is one of two Polish contributions to the BRITE constellation along with the Lem satellite. It is named after Johannes Hevelius.

In December 4, 2014, **PROCYON** (Proximate Object Close flyby with Optical Navigation) which is an asteroid flyby space probe was launched together with Hayabusa 2. It was developed by University of Tokyo and JAXA. It is a small (70 kg, approx. 60 cm cube), low cost (¥500 million) spacecraft.

On February 11, 2015, **Deep Space Climate Observatory** (**DSCOVR**; formerly known as **Triana**, unofficially known as **GoreSat**) which is a NOAA space weather and Earth observation satellite was launched by SpaceX on a Falcon 9 launch vehicle from Cape Canaveral.

On September 28, 2015, **Astrosat,** which is the India's first dedicated multi-wavelength space observatory was launched. After the success of the satellite-borne Indian X-ray Astronomy Experiment (IXAE), which was launched in 1996, the Indian Space Research Organization (ISRO) approved further development for a full-fledged astronomy satellite, Astrosat, in 2004.

On October 16, 2016, the **ExoMars Trace Gas Orbiter** (**TGO**) which is a collaborative project between the ESA and Roscosmos sent an atmospheric research orbiter and the *Schiaparelli* demonstration lander to Mars as part of the European-led ExoMars program.

The Trace Gas Orbiter delivered the *Schiaparelli* lander which crashed on the surface. The orbiter is currently aerobraking to gradually lower itself to a circular, 400 km (250 miles) orbit, where it will start optimum atmospheric mapping in April 2018. A key goal is to gain a better understanding of methane (CH_4) and other trace gases present in the Martian atmosphere that could be evidence for possible biological activity. The program will follow with the Surface Science Platform and the ExoMars rover in 2020, which will search for biomolecules and biosignatures; the TGO will operate as the communication link for the 2020 ExoMars rover and the Surface Science Platform and provide communication for other Mars surface probes with Earth.

3.2.5 Communication Satellites

It is only fair to start this section of the book with the person who is considered to be the godfather of satellite communication. In October of 1945, an RAF electronics officer who was a member of the British Interplanetary Society, **Arthur C. Clarke**, wrote a short article in Wireless World entitled "Extra Terrestrial Relays: Can Rocket Stations Give World-Wide Radio Coverage?"[56] This article is presented at the end of this section in its entirety because of its importance and relevance in our discussion. In his article, Clarke outlined the basic technical considerations involved in the concept of satellite communications. He proposed orbiting space stations that could be provided with receiving and transmitting equipment and could act as a repeater in space that would relay transmission between any two points of the hemisphere beneath. He calculated that an orbit with a radius of 42,000 km would coincide with the Earth's rotation, and the spacecraft would stay fixed as seen from any point on Earth. Clarke also explained that in this orbit, three satellites that are separated by 120° can provide worldwide communication coverage. It is interesting to note that Clarke did not patent his great idea because he thought that such a system of satellites would not be technically or economically feasible in the near future. This is when his clever theory came into reality in less than 20 years. In 1963 Clarke's idea became a reality with the successful launch of the first geosynchronous satellite SYNCOM by NASA. Some of the factors that were instrumental in the realization of satellite communication could be summarized as follows[57]:

- High-power rocket technology development and propulsion systems capable of delivering satellites to high-altitude orbits.
- Scientific and military interests in space research.
- Development of transistors and miniaturization of electronic circuitry.
- Development of solar cells for providing the source of energy for the satellite subsystems.
- High-speed computers development that could be used for calculating and tracking orbits.
- Government support for space technology.
- Space race between Soviets and Americans.

Clarke in his masterpiece summarized the advantages of what he called a space station as follows:

It is the only way in which true world coverage can be achieved for all possible types of services.

It permits unrestricted use of a band at least 100,000 Mc/s wide, and with the use of beams an almost unlimited number of channels would be available.

The power requirements are extremely small since the efficiency of "illumination" will be almost 100%. Moreover, the cost of power would be very low.

However great the initial expense, it would only be a fraction of that required for the world.

Networks replaced, and the running cost would be incomparably less.

One could ask why communication satellites? While there are so many other transmission systems that could relay telecommunication signals from one point to another. Such systems could be microwave, cable, troposcatter, fiber optics, and even a twisted pair of wires. The response would be that communication satellites are unique because they can provide global coverage with only three satellites at the Geostationary Earth Orbit, GEO. The unique characteristic of the GEO is that any object on this orbit is considered stationary with respect to the Earth because it will travel at the same speed and in the same direction as that of the Earth. Earth Station antennas will therefore need no tracking to receive signals from such satellites. Communication satellites can cover the remote, isolated, and low population areas and locations that could be extremely difficult if not impossible to reach otherwise. The installations and commissioning of the satellite systems are rather quick and could be done in a very short period of time. Another characteristic of communication satellite is that it has a high potential application in developing countries with insufficient terrestrial coverage. In some developing countries, the quality of terrestrial communication is not sufficient for services that need a reliable system that has a high rate of availability as well. The use of satellite system in cases such as the financial transactions that need a reliable system has been on the rise for these countries. One such system is VSAT (Very Small Aperture Terminal) system that is globally used and more so in developing countries as a reliable private network. Later in this section, more will be explained on the VSAT system. The satellite system could be also an alternative to the suboceanic cables. A rather recent satellite service, Digital Sound Broadcasting, by satellite has become an attractive alternative for CD quality radio service. Such service has become available since 2000, nationally and globally.

The satellite communication architecture is shown in Figure 3.46.

A satellite communication system is made up of three main segments. These three segments sufficiently describe the composition of the any given satellite system. These three segments are space segment, earth segment, and the link in-between. These three segments will be described in the following sections, and Figure 3.47 is a simplified demonstration of these three segments.

Communication satellites have a profound impact on our daily lives. From connecting the remote areas of the Earth and providing communication services around the country, to broadcasting the major events globally and saving lives during the natural disasters, satellite communication has found its way into everybody's life in so many different ways that the world without it would seem to be a very different world.

3.2.5.1 Satellite Orbits

A satellite system could be looked at as a repeater in the sky compared to the microwave system that relays signals terrestrially in the "line-of-sight." In addition to having a good knowledge of the three satellite segments, another factor that is essential in understanding the communication satellite system

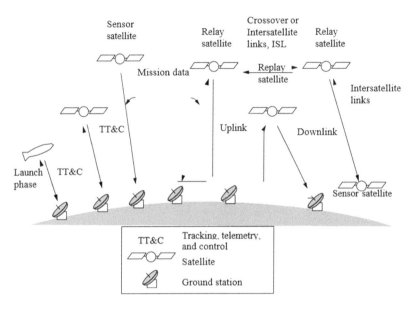

FIGURE 3.46
Satellite communications architecture.[58]

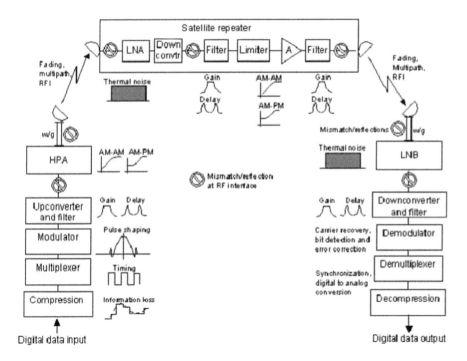

FIGURE 3.47
Communication satellite system.[59]

is the classes of satellite orbits. In answering a general question of why does a satellite stay in orbit, one could provide the following response. A satellite stays in orbit due to the balance of two effects: velocity, or the speed at which it would travel in a straight line; and the gravitational pull between the Earth and the satellite. This concept applies to all orbits used for satellite communications.

The following Figure 3.48 shows the types of orbits by inclination, by shape, and by altitude.

Figure 3.49 shows the launch speed required for a spacecraft to escape Earth's gravitational pull that differs.[60]

Briefly, the three main orbits used for commercial communication satellites have the following characteristics:

LEO (Low Earth Orbit)

    Minimize power in satellite

    Minimize power in handheld devices

    Minimize satellite antenna size

    Minimize time delay (latency)

    Located between atmosphere and the first Van Allen Belt (VAB)

    Maximize the angle of elevation

    Low cost of launch

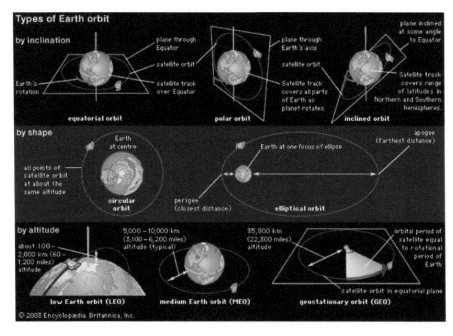

FIGURE 3.48
Basic characteristics of orbits in which a satellite can be placed around the Earth.[60]

Gravity versus launch speed

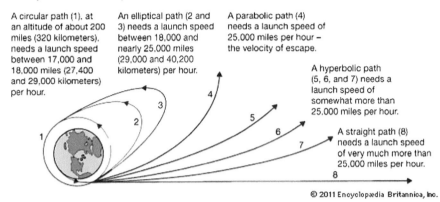

A circular path (1). at an altitude of about 200 miles (320 kilometers). needs a launch speed between 17,000 and 18,000 miles (27,400 and 29,000 kilometers) per hour.

An elliptical path (2 and 3) needs a launch speed between 18,000 and nearly 25,000 miles (29,000 and 40,200 kilometers) per hour.

A parabolic path (4) needs a launch speed of 25,000 miles per hour – the velocity of escape.

A hyperbolic path (5, 6, and 7) needs a launch speed of somewhat more than 25,000 miles per hour.

A straight path (8) needs a launch speed of very much more than 25,000 miles per hour.

© 2011 Encyclopædia Britannica, Inc.

FIGURE 3.49
The launch speed required for a spacecraft to escape Earth's gravitational pull differs.[60]

  Life in orbit of about 8 years

  Orbit: 400–1,600 miles

  Planet coverage: 48 satellites

  Time to complete one orbit: 100 minutes

 MEO (Medium Earth Orbit)

  Modest satellite antenna size

  Modest satellite power

  Small latency

  Less number of satellites than in LEO (around ten satellites depending on the exact orbit)

  Orbit: 1,500–6,500 miles

  Life in orbit: 6–12 years

  More complex tracking system than GEO

  Located between the two Van Allen Belts

 GEO (Geostationary Earth Orbit)

  Orbital location: 22,300 miles above the equator

  Time to complete one orbit: 24 hours

  Planet coverage: three satellites

  Signal delay time: 0.25 ms (each way)

  Life in orbit: 15–25 years

GEO commercial communication satellites are shown in Figures 3.50 and 3.51.[61]

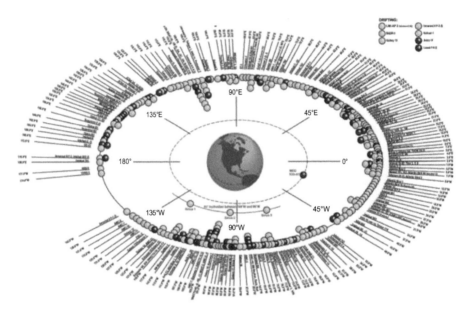

FIGURE 3.50
Commercial communication satellites in GEO. (Keesee, John, Satellite Communications.)

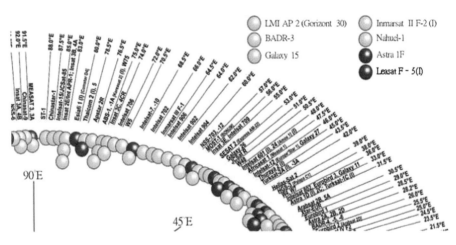

FIGURE 3.51
Segment of the GEO shown in Figure 3.50.

Some of the less commonly used orbits are as follows:

Polar Orbit (PO)

This orbit refers to spacecraft at near-polar inclination and an altitude of 700–800 km. The satellite passes over the equator and each latitude on the Earth's surface at the same local time each day, meaning that the satellite is overhead at essentially the same time throughout all seasons of the year. This feature enables collection of data at regular intervals and consistent times, which is especially useful for making long-term comparisons.

Highly Elliptical Orbits (HEOs)
These orbits are characterized by a relatively low altitude perigee and an extremely high-altitude apogee. These extremely elongated orbits have the advantage of long dwell times at a point in the sky; visibility near apogee can exceed 12 hours. These elliptical orbits are useful for communications satellites.

Geo Transfer Orbit (GTO)
This is an elliptical orbit of the Earth, with the perigee in LEO and the apogee in GEO. This orbit is generally a transfer path after launch to LEO by launch vehicles carrying a payload to GEO.

The other main parameter that is very important in understanding communication satellites operation is the frequency bands allocated to and used by communication satellite system. The ITU allocates frequency bands to all radio services and as such satellites have designated frequency bands that are being allocated in the World Administrative Radio Conferences (WARC). These allocations are expanded when the member countries demand such a need and the Conference approves it. The three major frequency bands used in communication satellites are C-band (6 GHz uplink/4 GHz downlink), K_u-band (14 GHz uplink/12 GHz downlink), and K_a-band (30 GHz uplink/20 GHz downlink). Table 3.15 shows the frequency band designations, their nominal frequency range, and principle applications. For allocation of

TABLE 3.15

Frequency Bands Designations

Band Designation	Nominal Frequency Range	Principal Uses
HF	3–30 MHz	Short-wave broadcast
VHF	30–300 MHz	FM, TV
UHF	300–3,000 MHz	TV, LAN, cellular, GPS
L	1–2 GHz	Radar, geostationary satellite orbit (GSO) satellites
S	2–4 GHz	Same as above
C	4–8 GHz	Satellite data links
X	8–12 GHz	Same as above
K_u	12–18 GHz	Radar, satellite data links
K	28–27 GHz	Same as above
K_a	27–40 GHz	Radar automotive data
mm (millimeter) wave	40–300 GHz	

frequencies, the world has been divided into three regions by the ITU as shown in Figure 3.52.

Figure 3.53 shows orbits of different satellites. Figure 3.54 shows the frequency allocation to different services. Figure 3.55 shows the frequencies used in satellite communications for commercial as well as government/ military services.

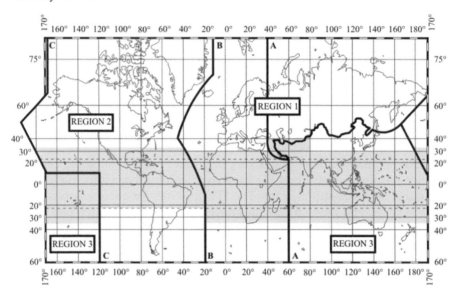

FIGURE 3.52
Three regions of the world according to the ITU for frequency allocation.[62,63]

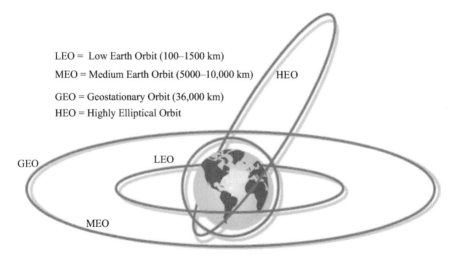

LEO = Low Earth Orbit (100–1500 km)
MEO = Medium Earth Orbit (5000–10,000 km)
GEO = Geostationary Orbit (36,000 km)
HEO = Highly Elliptical Orbit

FIGURE 3.53
Satellite orbits.

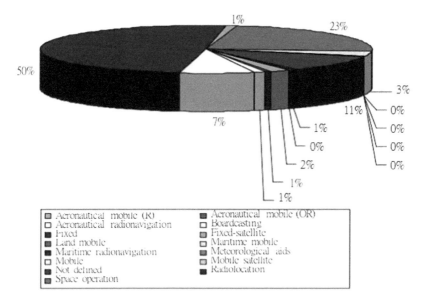

FIGURE 3.54
Frequency band allocation to different services.

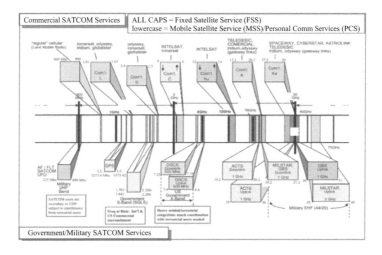

FIGURE 3.55
Satellite communication frequency usage. (Introduction to International Radio Regulations, ANNEX, Article 5 of the Radio Regulations, Edition 2001.)

Table 3.16 shows the C-band and K_u-Band frequency allocations to communication satellites for the above three regions of the world.[64]

The following tables show the frequency band allocations to satellite services in the upper K_u-band through the K_a-band, as listed in Article 5 of the RR.[65]

TABLE 3.16

ITU Frequency Allocation Table

Region 1	Region 2	Region 3
4500–4800	Fixed Fixed-Satellite (space-to-Earth) 5.441 Mobile	
6700–7075	Fixed Fixed-Satellite (Earth-to-space) (space-to-Earth) 5.441 Mobile 5.458 5.458A 5.458B 5.458C	
10.7–11.7	10.7-11.7	10.7-11.7
Fixed	Fixed	Fixed

Fixed-Satellite
  (space-to-Earth) 5.441
	Fixed-Satellite (space-to-Earth)
	  5.441
		Fixed-Satellite
		  (space-to-Earth) 5.441

5.484A (Earth-to-space) 5.484	5.484A	5.484A

Mobile except aeronautical
  mobile
	Mobile except aeronautical
	  mobile
		Mobile except aeronautical
		  mobile

12.75–13.25	Fixed	
	Fixed-Satellite (Earth-to-space) 5.441	
	Mobile	
	Space research (deep space) (space-to-Earth)	
	14.25 to 15.63 GHz	
Allocation to Services		

14.25–14.3	Fixed-Satellite (Earth-to-space)
	  5.484A 5.506
	Radionavigation 5.504
	Mobile-satellite (Earth-to-space)
	  except aeronautical mobile-
	  satellite space research 5.505
	  5.508 5.509	

14.4–14.4	14.3–14.4	14.3–14.4
Fixed	Fixed-Satellite	Fixed

Fixed-Satellite
  (Earth-to-space) 5.484A
  5.506
	Mobile-satellite (Earth-to-space)
	  except aeronautical mobile
	  satellite
		Fixed-Satellite
		  (Earth-to-space) 5.484A
		  5.506

Mobile except aeronautical
  mobile
	Radionavigational satellite	Mobile except aeronautical
		  mobile

Mobile-satellite
  (Earth-to-space) except
  aeronautical mobile
  satellite
		Mobile-satellite
		  (Earth-to-space) except
		  aeronautical mobile
		  satellite

Radionavigation-satellite		Radionavigation satellite

(Continued)

TABLE 3.16 (*Continued*)

ITU Frequency Allocation Table

Region 1	Region 2	Region 3

14.4–14.47	Fixed
	Fixed-Satellite (Earth-to-space)
	  5.484A 5.506 Mobile except
	  aeronautical mobile
	Mobile-satellite (Earth-to-space)
	  except aeronautical mobile
	  satellite
	Radionavigation-satellite Space
	  research (space-to-Earth)	

14.47–14.5	Fixed
	Fixed-Satellite (Earth-to-space)
	  5.484A 5.506 Mobile except
	  aeronautical mobile
	Mobile-satellite (Earth-to-space)
	  except aeronautical mobile
	  satellite
	Radio astronomy 5.149	

14.5–14.8	Fixed Fixed-Satellite (Earth-to-space) 5.510 Mobile Space research	
14.8–15.35	Fixed Mobile Space research 5.339	

15.35–15.4	Earth Exploration-Satellite
	  (passive)
	Radio Astronomy
	Space Research (passive) 5.340
	  5.511	

15.4–15.43	Aeronautical Radionavigation 5.511D	
15.43–15.63	Fixed Satellite (Earth-to-space) 5.511A Aeronautical Radionavigation 5.511C	

Source: International Telecommunications Union (ITU) Frequency Allocation Table, SMS4DC Training Seminar, October 2010.

Now that we have established this starting point in discussing the communication satellites, their orbits, and their frequency bands, let us look at some definitions before more details on satellite communication are discussed. This section covers the three main communication satellite services: FSS, MSS, and BSS.

Telecommunication—Any transmission, emission, or reception of signs, signals, writings, images, and sounds of intelligence of any nature by wire, radio, optical, or other electromagnetic systems.

Radio communication—Telecommunication by means of radio waves.

Space radio communication—Any radio communication involving the use of one or more space stations or the use of one or more reflecting satellites or other objects in space.

Radio-communication service—A service involving the transmission, emission, or reception of radio waves for specific telecommunication purposes.

Fixed service—A radio-communication service between specified fixed points.

FSS—A radio-communication service between earth stations at given positions, when one or more satellites are used; the given position may be a specified fixed point or any fixed point within specified areas; in some cases, this service includes satellite-to-satellite links, which may also be operated in the inter-satellite service; the FSS may also include feeder links for other space radio-communication services.

Inter-satellite service—A radio-communication service providing links between artificial satellites.

Space operation service—A radio-communication service concerned exclusively with the operation of spacecraft, in particular, space tracking, space telemetry, and space telecommand. These functions will normally be provided within the service in which the space station is operating.

Mobile service—A radio-communication service between mobile and land stations, or between mobile stations.

MSS—A radio-communication service.

Between mobile Earth stations and one or more space stations, or between space stations used by this service

Between mobile Earth stations by means of one or more space stations. This service may also include feeder links necessary for its operation

Land mobile service—A mobile service between base stations and land mobile stations, or between land mobile stations.

Land MSS—A MSS in which mobile earth stations are located on land.

Maritime mobile service—A mobile service between coast stations and ship stations, or between ship stations, or between associated onboard communication stations; survival craft stations, and emergency position-indicating radio-beacon stations may also participate in this service.

Maritime MSS—A MSS in which mobile Earth stations are located on board ships; survival craft stations and emergency position-indicating radio-beacon stations may also participate in this service.

Port operations service—A maritime mobile service in or near a port, between coast stations and ship stations, or between ship stations, in which messages are restricted to those relating to the operational handling, the movement and the safety of ships, and, in emergency, to the safety of persons. Messages that are of a public correspondence nature shall be excluded from this service.

Ship movement service—A safety service in the maritime mobile service other than a port operations service, between coast stations and ship stations, or between ship stations, in which messages are restricted to those relating to the movement of ships. Messages that are of a public correspondence nature shall be excluded from this service.

Aeronautical mobile service—A mobile service between aeronautical stations and aircraft stations, or between aircraft stations, in which survival craft stations may participate; emergency position-indicating radio-beacon stations may also participate in this service on designated distress and emergency frequencies.

Aeronautical MSS—A MSS in which mobile Earth stations are located on board aircraft; survival craft stations and emergency position-indicating radio-beacon stations may also participate in this service.

Broadcasting service—A radio communication service in which the transmissions are intended for direct reception by the general public. This service may include sound transmissions, television transmissions, or other types of transmission.

Broadcasting satellite service (BSS)—A radio communication service in which signals transmitted or retransmitted by space stations are intended for direct reception by the general public. In the broadcasting-satellite service, the term "direct reception" shall encompass both individual reception and community reception.

Before we discuss the three main services referred to above, it would be worthwhile to mention some of the highlights in the area of satellite communications during the last seven decades. Some of the selected communication satellites major milestones are as follows[35]:

1945—Arthur C. Clarke article: "Extra-Terrestrial Relays"

1955—John R. Pierce article: "Orbital Radio Relays"

1956—First Trans-Atlantic Telephone Cable: TAT-1

1957—Sputnik: Russia launches the first earth satellite.

1960—First successful DELTA launch vehicle

1960—AT&T applies to FCC for experimental satellite communications license

1961—Formal start of TELSTAR, RELAY, and SYNCOM programs

1962—TELSTAR and RELAY launched

1962—Communications Satellite Act (U.S.)

1963—SYNCOM launched

1964—INTELSAT formed

1965—COMSAT's EARLY BIRD: First commercial communications satellite

1969—INTELSAT-III series provides global coverage

1972—ANIK: First domestic communications satellite (Canada)

1974—WESTAR: First U.S. domestic communications satellite

1975—INTELSAT-IVA: First use of dual-polarization

1975—RCA SATCOM: First operational body-stabilized comm. satellite

1976—MARISAT: First mobile communications satellite

1976—PALAPA: Third country (Indonesia) to launch domestic communication satellite

1979—INMARSAT formed

1988—TAT-8: First fiber-optic trans-Atlantic telephone cable
     First land mobile satellite system

1990—Badr-1: First communications satellite of Pakistan

1991—ITALSAT (ITALY): The first satellite with onboard processing and multibeam coverage

1994—Turksat 1B: First communications satellite for Turkey

1997—Iridium 1: First satellite for satellite telephone service

1998—Iridium phone service begins

2000—AO-40: First satellite to use momentum wheels with magnetic bearings
     First satellite to use GPS for navigation and attitude determination in HEO

2000—Globalstar data service begins

2000—Second-generation tracking and data relay satellite system (TDRSS) satellites launched

2001—Iridium 2.4 kbps data service begins

2002—Inmarsat F77 MPDS (by-the-byte) service begins

2006—SuitSat: First reuse of a decommissioned spacesuit as a radio satellite

2007—Inmarsat Fleet BroadBand service begins

2008—Significant 2008 events include:

Iridium OpenPort (128 bps) service begins

SES Astra's Astra-1M satellite (Luxembourg) was launched from the Baikonur Cosmodrome on November 6, 2008, by an International Launch Services (ILSs) Proton M rocket.

On February 11, a Proton M rocket launched AM44 and Express MD-1 for the Russian Satellite Communications Co. (RSCC).

On February 12, an Arianespace Ariane 5 ECA launched two communications satellites, Eutelsat's Hot Bird 10 and SES New Skies' NSS-9.

Telesat Canada's Telstar 11N satellite was successfully orbited on February 26 by a Land Launch Zenit-3SLB rocket from the Baikonur Cosmodrome, replacing Telstar 11, which was launched in 1995.

Ciel Satellite Group's Ciel-2 (Canada) was launched into a geosynchronous transfer orbit by an ILSs Proton Breeze M rocket from the Baikonur Cosmodrome on December 10, 2008.

On September 17, Telesat's Nimiq 5 satellite (Canada) was launched.

On October 1, an Ariane 5 ECA launched Hispasat's (Spain) Amazonas 2 spacecraft, built by EADS Astrium (Europe) on their Eurostar E3000 platform.

2009—Significant 2009 events include:

Launch of JAXA's Ibuki spacecraft by a Mitsubishi H-2A.

Launch of Japan's H2 Transfer Vehicle (HTV) to the International Space Station (ISS) by the first H-2B rocket.

Suborbital flight of NASA's Ares I-X test vehicle.

Four NASA space shuttle launches to the ISS.

Intelsat and SES spearheaded the creation of an industry-wide global database to collect information on interference and to endorse standardized training and certification for the personnel who install and operate satellite uplink ground stations.

The heaviest and the most powerful commercial satellites ever launched were orbited this year: the 6.91-tonne mobile-communication satellite TerreStar-1 and the 20-kW Sirius-FM5 radio broadcast satellite.

2010—Significant 2010 events include:

Take-off of the Latin American market.

The successful launch of Ka-Satellite, the Ka-band technology will be used within the continental United States in 2013 and expand to worldwide coverage in 2015. In the development of the satellite broadband market, Europe launched the KaSat and Hylas-1 satellites.

The Galaxy 15 rogue satellite.

Deepwater Horizon and the effects of oil and gas on satellite communications.

The rise of UAVs, and the return of broadband K_u-band satellite connectivity on commercial passenger airlines.

SpaceX successfully completing its second Falcon 9 launch and recovering its Dragon capsule after orbiting the Earth.

2011

Combinations of FSS and MSS spectrum ownership and operations.

Reintegration of satellite and terrestrial telecom operations, if not of ownership.

EchoStar announcing in February an agreement to acquire Hughes.

2012

Navid (Persian: نوید) or **Navid-e Elm-o San'at** ("Gospel of Science and Technology") was an experimental Iranian Earth observation satellite. The satellite carried a camera for taking higher resolution imagery of Earth and it was also be used to collect weather data and monitor natural disasters. It was developed by students at the Iran University of Science and Technology. The third satellite to be launched indigenously by Iran, it was placed into orbit by a new configuration of the Safir carrier rocket, featuring a larger second stage with 20% more thrust. The launch occurred at approximately 00:04 UTC on February 3, 2012. The satellite remained in orbit for 2 months, before reentering the atmosphere on April 1, 2012.

2013

CASSIOPE, or *CAScade, Smallsat and IOnospheric Polar Explorer*, is a Canadian Space Agency (CSA) multi-mission satellite operated by MacDonald, Dettwiler and Associates (MDA). The mission is funded through CSA and the Technology Partnerships Canada program. It was launched September 29, 2013, on the first flight of the SpaceX Falcon 9 v1.1 launch vehicle. CASSIOPE is the first Canadian hybrid satellite to carry a dual mission in the fields of telecommunications and scientific research. The main objectives are to gather information to better understand the science of space weather, while verifying high-speed communications concepts through the use of advanced space technologies.

2016

The **Maritime Monitoring and Messaging Microsatellite** (M3MSat) is a tele-detection satellite developed by the CSA and launched in 2016. Its mission is to demonstrate and test the technology to assess the utility of having in space an Automatic Identification System (AIS) for reading signals from vessels to better manage marine transport in Canadian waters. The system will be supported by an instrument called a Low Data Rate Service (LDRS), which transmits AIS messages to ground sensors.

3.2.5.2 Satellite Structure

Although each satellite varies in shape, size, and mission, they all have something in common. Each satellite needs a power source, has instrumentation consisting of scientific and engineering sensors that measures changes in the satellite and its surroundings, and has special propulsion systems aboard called thrusters to push the satellite into its desired orbit. Small thrusters provide attitude, altitude, and propulsion control to modify and stabilize the satellite's position in space. Guidance and control sensors keep the satellite on its proper course, and sensors such as horizon seekers, star trackers, and sun seekers help determine the satellite's positions. Satellites are permitted to move from their allocated orbital location within specified limits, i.e., 0.01 deg. If the satellite does deviate from this limit which will result in adjacent interference into the neighboring satellite, then the thrusters will be activated through a command from the TT&C station by an engineer to move the satellite back to its allocated position.

The major components of any man-made satellite are as follows:

- Communications capabilities with Earth
- A power source
- A control system to accomplish its mission

The design of the spacecraft is a complicated exercise involving just about every branch of engineering and physics. Communications performance requirements, the need for a benign environment for communication equipment, and the process of launching into the desired orbit are among the subjects of space systems engineering.

The structure to hold the spacecraft together must be designed to withstand a variety of loads. During the launch of satellites, there are numerous factors that are essential for a successful launch. Some of these parameters are accelerations, vibration, aerodynamic loads, centrifugal stresses, operating thrust, and separation shocks. Once in orbit, again the operating thrusts, centrifugal stresses, thermal stresses, and exposure to charge particle radiations are of critical concern, and they need to be carefully monitored. Material selection for spacecraft structure design requires very careful consideration based upon material strength, thermal properties, electrical properties, strength, stiffness, toxicity, and shielding ability. A wide variety of materials and techniques have been used for spacecraft structure. The following table[56] lists some common structural materials (Table 3.17).

TABLE 3.17

Lists of Some Common Structural Materials

Aluminum
Magnesium
Stainless steel
Invar
Titanium
Graphite-reinforced phenolic
Fiberglass epoxy
Beryllium

Figure 3.56 shows an example of a satellite bus and payload structure.

Applied Aerospace Structures Corporation (Stockton, California) has fabricated over 120 satellite structures from 1968 through 2007 and has established industry-recognized manufacturing expertise because of their extensive history of flight-proven composite and aluminum/honeycomb structures. These structures include bus, payload, and instrument structures, a sample of which is shown in Figure 3.56.

3.2.5.2.1 Attitude and Orbit Control SubSystem

- The attitude of a satellite is the direction at which the satellite points to with respect to Earth. Communication satellites, and most other

FIGURE 3.56
Satellite bus and payload structure.

types of satellites, must point to Earth or point in some other specific direction. A communication satellite must point towards Earth
because it has high gain antennas (directional antennas that must
be pointing toward the Transmitter/Receiver to be able to receive/
transmit). It is very important that the direction to which a satellite
is pointing is knows and controlled.

• It would be a huge waste to use rocket fuel to control the satellite attitude because this is done very often within the course of a single orbit.
So other means of turning the satellite are used. Satellites are usually
equipped with what is known as momentum wheels to rotate the satellite about its axis. When the momentum is spun in a specific direction about some axis, the satellite will spin the opposite direction.

Some satellites need to be stabilized in several dimensions while others need
to be stabilized in only a single dimension:

3-axis stabilized satellites: contain three momentum wheels that
are fixed in three dimensions, one for each dimension, or a single
momentum wheel that is installed on gimbals is used instead. This
type of satellites has to be stabilized in three dimensions for its
antennas to face Earth and its solar cells to face the Sun for maximum power generation.

Spinner satellites: are stabilized only in one dimension. The other two
dimensions are stabilized by having the satellite spin around an axis
at rates from 30 to 60 rotations per minute. In this case, satellite antennas must either have circular symmetry such as omnidirectional
antennas (e.g., monopoles) or must be de-spun (antennas are rotated
at an opposite direction to the satellite rotation) for the antennas to
appear to be stationary. Bearings are used in this process. Satellite
orbital control is achieved by rocket engines that use fuels such as
hydrazine (N_2H_4) or arc jets (ion thrusters). Orbital control is very
important for GEO satellites to prevent loss of signal at Earth stations.

3.2.5.2.2 Telemetry, Tracking, Command, and Monitoring (TTC and M) Subsystem

Telemetry and Monitoring: is the process of measuring different parameters of the satellite and determining the condition of a satellite and its health,
which include hundreds of parameters, obtained from sensors onboard the
satellite and sending them back to the Earth station through the telemetry
signal to take actions. Some of these parameters include:

• Pressure in fuel tank indicating amount of fuel remaining,
• Power provided by solar cells,
• Power consumed by different communication devices,

- Temperature of different parts of the satellite,
- Position of switches that control different devices on the satellite,
- Attitude information

Tracking: is the process of determining exactly where the satellite is located, where it is heading, its speed, and its acceleration.

- **Velocity and Acceleration Measurements:** Velocity may be measure by integrating acceleration, which may be obtained using accelerometers (acceleration sensors), and then adding the initial value of the velocity to get a very good estimate of the satellite velocity at any time,
- **Beacon Signal Doppler Shift:** Satellites usually have beacons (transmitters that transmit signals with specific know frequencies). Earth stations monitor the Doppler shift of beacon transmissions to determine the speed of the satellite.
- **High Gain Antenna Tracking:** Earth stations usually use very high gain antennas that have very narrow beam-widths. So, the position of a satellite can be determined by pointing a very high gain antenna toward it and moving the direction of the antenna to track the satellite movements.
- **Echoing Delay Measurement:** If the processing delay of a satellite is known, Earth stations can monitor the pulse echoing delay echoing delay to determine how far a satellite is from the Earth station.

Command: is the process of ordering the satellite to perform some operation.

- **Command Channel:** to avoid having any unauthorized tampering with the satellite, secure channels (encryption of commands and responses) are used for sending commands to the satellite and receiving responses.
- **Types of Commands:** The usual commands to a satellite may be to:
  - Fire its rocket thrusters to correct its orbit,
  - Correct its attitude,
  - Shutdown or turn on some parts of its communication systems,
  - Change the position of some switches,
  - Extend solar cells,
- **Redundancy of Commands:** to avoid having any command being misinterpreted by the satellite, which may cause a disaster, the commands are echoed back the Earth station before being executed, and the execution requests are also echoed back to the Earth station after being executed.

3.2.5.2.3 *Power SubSystem*

Batteries: Because satellites sometimes pass through the shadow of Earth for it to remain operational, satellites are equipped with batteries to store excess energy from the solar cells during the period in which the solar cells are illuminated by the Sun and provide the satellite with energy when satellite is in the dark:

The number of batteries needed should be sufficient to support the operation of a GEO satellite for the maximum eclipse duration of 70 minutes every day. The longest eclipses occur in the Fall and Spring (around 21 September and 21 March of each year). Shorter eclipse periods occur before and after these dates. During a big part of the year, no eclipses will occur.

If batteries cannot support the whole satellite, parts of the communication system of a satellite may be shut down for the whole period of eclipse and then started again once the satellite goes out of Earth's shadow.

Batteries used on Satellites usually have ratings of 20–50 V and 20–100 Ah.

The efficiency of batteries deteriorates with time, so when designing the satellite, the amount of batteries to be included is usually based on the end of life efficiency of the batteries. That is, how much energy would the batteries be able to store near the end of life of the satellite.

Solar cells are the element that provides the whole communication satellites with power to perform all of its tasks. The sun provides an amount of energy equal to approximately 1.361 kW of power per square meter at Earth's distance from the sun. Although Earth's atmosphere absorbs a significant amount of this power, since satellites are in space, this is the amount of energy that they can theoretically get from the sun per square meter. The best solar cells in the market can only extract a fraction of the incident solar power and convert it to electric power. Typically, the best solar cells have an efficiency of only 20–25%. Solar cells may have efficiencies as high as 25% when they are first manufactured. However, with time, this efficiency drops because of many factors including aging and the small scratches that occur because of small meteors that constantly hit them. Therefore, it is expected that the efficiency of a solar cell may drop from 25% to 15%, for example, at the end of the satellite life. Therefore, when building a satellite, it makes sense to have extra solar cell area to compensate for the expected drop in efficiency near the end of life of the satellite. Solar cells convert solar power to electricity. The more solar light a solar cell blocks the more the electric power that will be generated. Therefore, solar cells that are directed such that solar rays hit them perpendicular to their plane will generate the maximum amount of power. If the incident light and the line perpendicular to the solar cell are parallel, the solar cell will produce the maximum amount of power. If the angle between these lines is θ, the amount of converted power will be equal to a fraction of cos (θ) of the maximum power. If solar light is parallel to the surface of the solar cell (i.e. $\theta = 90°$), the produced power will be 0. Communication satellites typically need an amount of power from 5 to 10 kW.

It is to be noted that the 3-axis stabilized satellites have flat solar cell panels that are fixed on rotating shafts to allow the satellite to always point its solar cell panels directly to the sun. Spin-stabilized satellites, however, have their solar cells covering its cylindrical surface, so solar cells have a cylindrical shape and only one half of the total area of solar cells will be illuminated by the sun. In addition, not all of the illuminated region will produce maximum power. Assuming that the solar cells on a satellite are directed as to generate the maximum power, if the total area of solar cells on the satellite is A m² and they have efficiency η; then:

$$P_{3\text{-axis stabilized}} = (1{,}361 \ \eta \ A) \ W$$

$$P_{\text{spin-stabilized}} = (1{,}361 \ \eta \ A)/\pi \ W$$

Because solar cells of spinner satellites are not fully exposed to solar power at all time, they remain at a lower temperature than solar cells of 3-axis stabilized satellites. The lower temperature of solar cells usually results in a higher efficiency. So, the efficiency of solar cells of spinner satellites is usually higher than the efficiency of solar cells of 3-axis stabilized satellites.

3.2.5.2.4 Satellite Structural Design

The cost of launching a satellite is a function of its mass. As a result of this, the cost of launching one is very high, more so in the case of geostationary satellite. One of the most basic requirements is therefore lightness of its mechanical structure. All efforts are therefore made to reduce the structural mass of the satellite to the minimum. This is achieved by using materials that are light yet strong. Some of the materials used in the structure include aluminum alloys, magnesium, titanium, beryllium, Kevlar fibers, as mentioned above and more commonly the composite materials. The design of the structural subsystem relies heavily on the results of a large number of computer simulations where the structural design is subjected to stresses and strains similar to those likely to be encountered by the satellite during the mission. The structural subsystem design should be such that it can withstand mechanical accelerations and vibrations, which are particularly severe during the launch phase. Therefore, the material should be such that it can dampen vibrations. Kevlar has these properties. The satellite structure is also subjected to thermal cycles throughout its lifetime. It is subjected to large differences in temperature as the sun is periodically eclipsed by Earth. The temperatures are typically several hundred degrees Celsius in the side facing the sun and several tens of degrees below zero degrees Celsius on the shaded side. Designers keep this in mind while choosing material for the structural subsystem.

The space environment generates many other potentially dangerous effects. The satellite must be protected from collision with micrometeorites, space junk, and charged particles floating in space. The material used to cover the outside of a satellite should also be resistant to puncture by these fast travelling particles. The structural subsystem also plays an important role in ensuring reliable operation in space of certain processes such as separation of the satellite from the launcher, deployment and orientation of solar panels, precise pointing of satellite antenna, operation of rotating parts, and so on.

A typical satellite consists of a number of vital subsystems, and of a payload carried for the ultimate mission purpose. To summarize the above discussion regarding the satellite structure we can say that a "subsystem" is a group of single components that are organized in working units. The usual subsystems that make a satellite working are as follows (Figure 3.57):

1. Structure and mechanisms: they carry the payload and keep all the other subsystems together. They are often the heaviest spacecraft hardware, so they affect a number of challenges like launch loads, material stability in vacuum and direct sunlight radiation, resistance to vibrations and shocks.

2. Electric power subsystem: every satellite needs energy, so it needs a power subsystem to generate, control, store, and distribute electrical current along every working component. This way, an electric power subsystem is often divided in four smaller parts, like a power

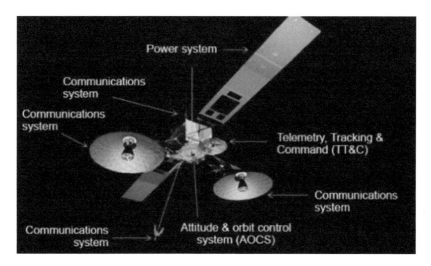

FIGURE 3.57
Satellite subsystems.

source (solar arrays), a power storage device (battery), a power control station, and a power distribution structure. The electrical components must also be qualified for vacuum and solar radiation operations.

3. Thermal control subsystem: as a satellite's core is frequently made of integrated electronic processors (the "thinking brain"), it needs to keep an adequate working temperature for all the units in some allowed ranges. The very different kind of solar exposition that a satellite usually faces, as all equipment is exposed to the longest direct sunlight during the day and on the other side is completely in darkness when behind Earth's shadow should be carefully looked into during the design process.

4. Attitude control subsystem: this subsystem is aimed to direct the satellite into desired directions and stabilize the satellite attitude.

5. Onboard data handling system: it controls the handling and the storage of satellite's health data and all the data generated by the (eventual) payload. The data collected by all of the sensors on board of the satellite to monitor the status of each subsystem to be relayed to the TT&C stations, are part of the data handling system.

6. Communication subsystem: to assure the ground-satellite communication in both uplink and downlink directions. Usually it consists of one or more receivers that can be deployed and oriented. Reliability is a primary issue within this specific subsystem, as it is the ultimate connection between the mission control center and the satellite in orbit. Transponders, antennas, frequency converters, etc. are parts of the Communication subsystem.

7. Payload: not always present, the payload is aimed to perform the mission objectives. For instance, a high-resolution camera constitutes the normal payload of an Earth imagery satellite.

8. Propulsion subsystem: the engine of a satellite, to perform orbit maneuvers and potentially change orbit's altitude or trajectory. It can be used to send the satellite into a re-entering orbit or to transfer broken spacecraft into what is known as "graveyard orbits," in order to avoid collisions with other spacecraft.[66]

Overall Satellite structure, Figure 3.58 is composed of space segment and ground segment. In this section, the focus is on the space segment and its subsystems.

The following chart shows the essential elements of the space segment, which are the bus and the payload. The payload is mainly composed of the transponders and antennas. The communication satellites' transponders are usually expressed in 36 MHz and sometimes in 72 MHz units. Transponder's

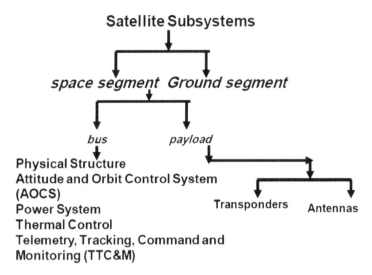

FIGURE 3.58
Overall satellite structure.

main function is to amplify the received signal, change its frequency, and transmit it. The change of frequency could be from the uplink frequency to downlink or it could be from one frequency band to another such as from C-band to K_u-band or vice versa.

As it is obvious, communication subsystem is the most important system of a communication satellite. All other subsystems are there to serve this subsystem. The core of the communication subsystem is *transponder*, which as mentioned above, receives a signal from an Earth station, changes its frequency, amplifies it, and then transmits it to another Satellite or back to Earth. The number of transponders in a Communication Satellite is typically 10–50 transponders depending on traffic of the covered region and other factors that will be addressed later in this chapter.

It is clear that the use of the same frequency for the uplink as well as the downlink would cause a big problem in the amplifiers. Even if very highly directional antennas were used in the satellite for the uplink and downlink, there would always be some signal leakage from the transmitting antenna on a satellite to the receiving antenna. Since the signals received by satellite are so weak, transponders use filters of very high gains. Even if a small amount of transmitted signal leaked to the receiving antenna, it would be highly amplified, which would cause even more leakage of the transmitted signal back to the receiving antenna. This would cause oscillation in the amplifier and it would saturate, making the amplifier useless. The only way to avoid this is to make sure that the transmitted signal has a different frequency from the received signal. The uplink/downlink frequencies of several frequency bands used in satellites are: C-band (6 GHz/4 GHz), K_u-band (14 GHz/12 GHz), and K_a-band (30 GHz/20 GHz). Transponders have different bandwidths depending on the band they operate at. Typical transponder bandwidths are 36 MHz (C-band), 54 MHz (K_u-band), and 72 MHz (K_a-band) as mentioned above. Knowing that the bandwidth of a standard definition TV channel is around 4 MHz, after it has been digitized and compressed, and the bandwidth of a voice signal is around 4 kHz, this means that a transponder may process from around 8 to 17 channels, at C-band or K_a-band, respectively, and additionally can process hundreds of voice channels as well.

Frequency Reuse in Satellites is another feature of the satellites, which uses techniques that allow a single satellite to reuse the allocated bandwidth several times. A satellite with bandwidth of 500 MHz, for example, may be able to effectively have a bandwidth of up to seven times by using the frequency reuse seven-fold. One of the methods used to achieve this frequency reuse is using two orthogonal polarizations.

Another feature of transponders is that they come in either single conversion or double conversion forms. In single conversion transponders, the frequency of the incoming signal (at Uplink RF frequency) is directly shifted to the frequency of the outgoing signal (at Downlink RF frequency). In double conversion transponders, the frequency of the incoming signal (at Uplink RF frequency) is first converted to a low intermediate frequency (IF frequency), then the frequency is shifted to the frequency of the outgoing signal (at Downlink RF frequency).

The typical power transmitted by a transponder is around 200 W or so. This is equivalent to around 20 W per TV channel.

The following shows a block diagram of a typical narrowband single conversion transponder. The bandwidths of the amplifiers and filters are approximate values of what may be found in a C-band transponder.

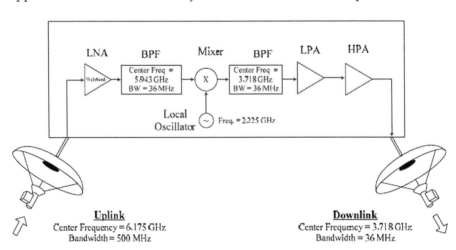

The following shows a block diagram of a typical narrowband double con-version transponder. The bandwidths of the amplifiers and filters are typical values that may be found in a C-band transponder.

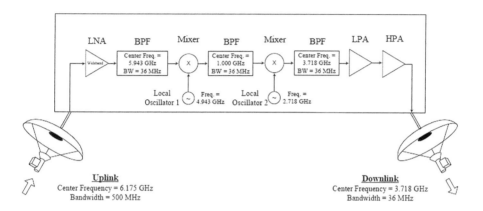

3.2.5.3 Frequency Bands and Allocations

The frequency spectrum is shown in the following figure:

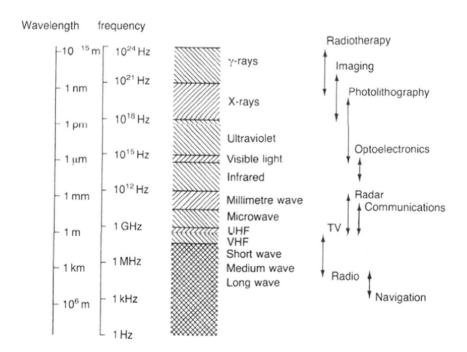

Frequency band allocation to communication satellites is a very complex and detailed process that takes place at ITU for global allocations and at the country's or region's organization that is so assigned to this task such as Federal Communications Commission (FCC) in USA. According to ITU, the world is divided into three radio regions as showed in the following figure. The frequency coordination is carried out globally in the WARC or regionally in the regional administrative radio conferences (RARC).

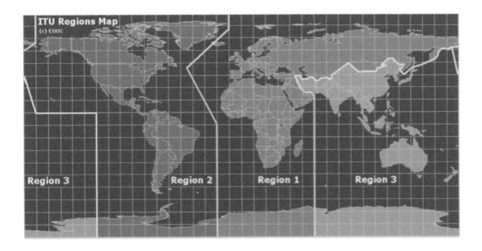

The ITU-R is the sector that is responsible for all global radio telecommunications services as shown in the broader picture of the ITU structure in the following figure:

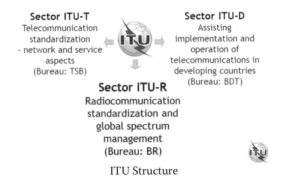

ITU Structure

Frequency band allocation to satellites is one of the many tasks carried out by ITU-R, the structure of which is shown below:

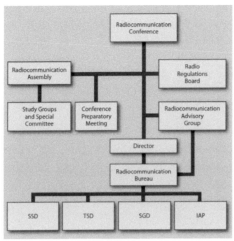

ITU-R Structure

The ITU-R Study Groups are assigned different topics as shown in the following figure:

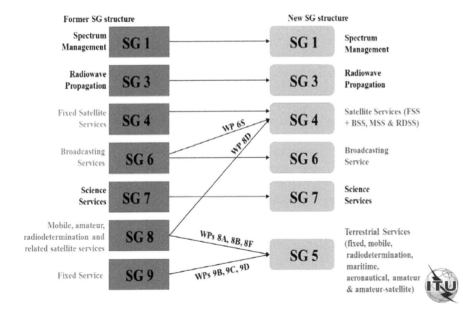

As it could be seen from the above figure, SG 4 was assigned the fixed satellite services and SG 8 was assigned the MSSs in the former SG structure and SG 4 with the task of FSS, BSS, and MSS services all combined in one group in the new study group structure. The frequency spectrum, in which the satellite/microwave telecommunication (1–40 GHz) is only a part of, is shown below.[67] The expanded section of the spectrum shown on the bottom of the following figure showed the detailed allocations of RF frequency bands in each of the designated band.

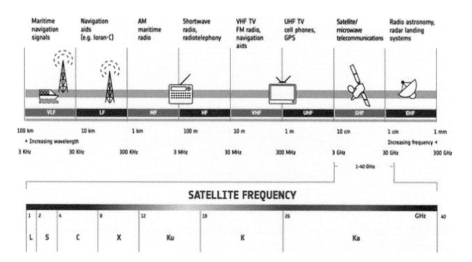

The three most frequently bands used in satellite communications are C-band, K_u-band, and K_a band. As it is explained in different sections of this book, there are some basic characteristics of these three frequency bands that are worth mentioning here in an integrated form. As the principle of radio propagation applies, the attenuation of radio signal increases with the increase of frequency. The higher the frequency, the more would be the signal attenuation. The basic frequency/wavelength relationship of $\lambda = c/f$ says that as the frequency increases, the wavelength becomes smaller and as the wavelength gets small enough to be comparable to the size of particles in air, the attention reaches its maximum and the signal gets absorbed, scattered, reflected, or refracted by these particles and the result is zero or close to zero of the signal reaches the receiving antenna.

There are other factors involved in relation to which frequency we operate at and to simplify these factors, the following table should be carefully studied:

f (GHz) Factors	C-Band	K_u-Band	K_a-Band
Attenuation	Small	Medium	Large
Interference[a]	Large	Small	No
Antenna size	Large	Medium	Small
Bandwidth, MHz	500	750	3,500

[a] Interference with terrestrial systems.

Satellite technology is developing fast, and the applications for satellite technology are increasing all the time. Not only can satellites be used for radio communications, but they are also used for astronomy, weather forecasting, broadcasting, mapping, and many more applications. With the variety of satellite frequency bands that can be used, designations have been developed so that they can be referred to easily. The higher frequency bands give access to wider bandwidths, but are also more susceptible to signal degradation due to "rain fade" (the absorption of radio signals by atmospheric rain, snow, or ice). Because of satellites' increased use, number, and size, congestion has become a serious issue in the lower frequency bands. New technologies are being investigated so that higher bands can be used.

L-Band (1–2 GHz): GPS carriers and also satellite mobile phones, such as Iridium; Inmarsat providing communications at sea, land and air; WorldSpace satellite radio which was the very first and still is the only global digital sound broadcasting satellite system.

S-band (2–4 GHz): Weather radar, surface ship radar, and some communications satellites, especially those of NASA for communication with ISS and Space Shuttle. In May 2009, Inmarsat and Solaris mobile (a joint venture between Eutelsat and Astra) were awarded each a 2×15 MHz portion of the S-band by the European Commission.

C-Band (4–8 GHz): Primarily used for satellite communications, for full-time satellite TV networks or raw satellite feeds. Commonly used in areas that are subject to tropical rainfall, since it is less susceptible to rainfade than K_u band (the original Telstar satellite had a transponder operating in this band, used to relay the first live transatlantic TV signal in 1962). This is the most commonly used frequency band in domestic, regional, and global communication satellites. Even today with the advancement of technology that has made K_u band available and to some extent the K_a band, still C band is the most used frequency band in satellite communication.

X-Band (8–12 GHz): Primarily used by the military. Used in radar applications including continuous-wave, pulsed, single-polarization, dual-polarization, SAR and phased arrays. X-band radar frequency sub-bands are used in civil, military, and government institutions for weather monitoring, air traffic control, maritime vessel traffic control, defense tracking, and vehicle speed detection for law enforcement.

K_u-Band (12–18 GHz): Used for satellite communications. In Europe, K_u-band downlink is used from 10.7 to 12.75 GHz for direct broadcast satellite services, such as Astra. Most of the communication satellites today are using a combination of K_u and C band transponders. Although K_u band provides more bandwidth than C band, but the attenuation due to rain, snow, ice, and in some areas, sand storm (especially when it is moist) is much higher and limits its use in severe weather conditions.

K_a-Band (26–40 GHz): Communications satellites, uplink in either 27.5 GHz or 31 GHz bands, and high-resolution, close-range targeting radars on military aircraft. As was mentioned before in this chapter, the higher the frequency band, the more attenuations will result. K_a band, although it provides much higher bandwidth than C band and K_u band but because of higher attenuation, its application is limited in some areas. It is to be noted again that the advantage of higher frequency is not only its higher bandwidth but smaller ground station antennas and no interference with terrestrial systems such as microwave stations, which operate at lower frequency bands.

3.2.5.4 Modes of RF Interference in FSS[68,69]

The quality of signals received by the satellite transponder and that retransmitted and received by the receiving earth station is important if successful information transfer via the satellite is to be achieved. RF interference has the effect of adding to the overall noise on the link, thereby increasing the receiving system noise temperature and degrading the quality of the received signal. The solution to this is to study potential sources, types, and levels of radio frequency interference (RFI) and their effect on system performance. Intersystem interference can be minimized by adjusting the antenna discrimination and receiver selectivity, and choosing proper frequency plans. The control of external interference partly depends on coordination, control, and sometimes compromising on the radio channel through direct negotiation. Earth stations must comply with carefully considered performance criteria in order to ensure that the various users of a satellite system can share the satellite resources without unduly affecting each other's service quality. The design, installation quality, and performance of an Earth station determine both the sensitivity of transmissions to interference from other services and the potential for the transmissions to interfere with other services. The common types of RF interference in fixed satellite systems are:

- Adjacent Satellite Interference (ASI).
- Co-Channel Interference (CCI).
- Cross-Polarization Interference (XPI).

3.2.5.4.1 Adjunct Satellite Interference

RF spectrum and the geosynchronous orbit are natural resources; therefore, geosynchronous satellites must share limited space and frequency spectrum within a given arc of a geostationary orbit. Each communication satellite is assigned a longitude in the geostationary arc approximately 35,786 km above the equator. The position in the slot depends on the communications frequency band used. Satellites operating at or near the same frequency must be separated in space to avoid mutual interference.

Generally, spatial separation of 1° to 4° is required depending on the following variables:

- Beamwidth and side lobe radiation of both the Earth station and satellite antennas.

- RF carrier frequency
- Encoding or modulation technique used
- Acceptable limits of interference
- Transmit carrier power

ITU mandates that at least 2° separation between adjacent fixed satellites, although countries can reduce this spacing on their allocated arc. United States, for example, uses 1° separation between satellites since there is more restrictive design specifications on the satellites and Earth station antennas. Of course, the closer the satellites are, assuring no interference with the neighboring satellites, the higher is the efficiency in the use of the GEO which is a limited natural resource, as was mentioned earlier.

A transmit Earth station can inadvertently direct a proportion of its radiated power toward satellites that are operating at orbital positions adjacent to that of the wanted satellite. This can occur because the transmit antenna is badly pointed toward the wanted satellite, or because the Earth station antenna beam is not sufficiently concentrated in the direction of the wanted satellite.

This unintended radiation can interfere with services that use the same or similar frequencies on the adjacent satellites. Similarly, a receiving Earth station can inadvertently receive transmissions from adjacent satellite systems which then interfere with the wanted signal. This happens because the receiving antenna, although being very sensitive to signals coming from the direction of the wanted satellite, is also sensitive to transmissions coming from other directions. Figure 3.59 shows two neighboring geosynchronous satellites and interferences caused due to each other.

If a transmitting Earth station (Earth station 1) radiates a small percentage of its EIRP toward another satellite (satellite B), this interfering signal appears as noise in the information bandwidth. The noise signal in a frequency translating transponder is amplified and re-radiated to the receiving Earth station 3. As a result, the re-radiated interfering downlink and the wanted downlink signal combine and appear as noise in the receiving Earth station 3. Similarly, if satellite B radiates some interfering signal to Earth station 2, then this interfering signal combines with a wanted downlink signal from satellite A. Although antennas used in satellite communications use a highly focused beam, exactly how focused the main beam is, or the reduction in transmitted (or received) power at any angle off the main beam, is determined by equipment design or, more specifically, by a particular antenna discrimination pattern.

3.2.5.4.1.1 Angular Separation   The angular separation between satellites, as seen by the Earth stations, influences the level of interference generated or received from the side lobe of the Earth station antenna into or from an adjacent satellite. Side lobe characteristic is one of the main factors in determining

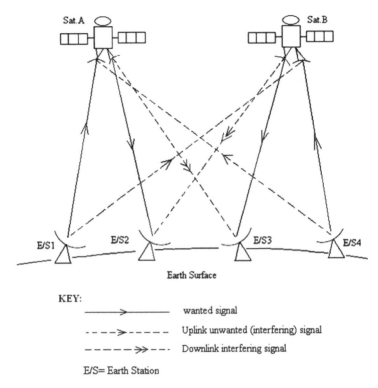

KEY:

→ wanted signal

- - -≫- - - - Uplink unwanted (interfering) signal

- - - - -≫- - - Downlink interfering signal

E/S= Earth Station

FIGURE 3.59
Two neighboring geosynchronous satellites and interferences due to each other.

the minimum spacing between satellites and, therefore, the orbit/spectrum efficiency. Figure 3.60 shows a satellite broadcast receiver with an antenna front-end that receives the wanted signal from satellite 1 with interferers on adjacent satellites 2 and 3 due to insufficient angular separation.

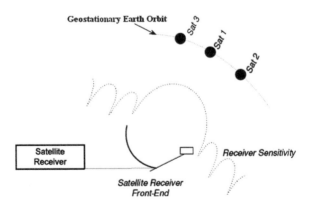

FIGURE 3.60
Satellite receiver receiving wanted signal with interferers.

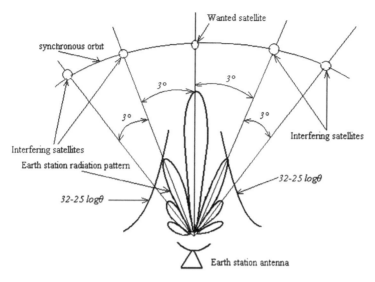

FIGURE 3.61
Adjacent satellite interference.

Figure 3.61 shows two "unwanted" or interfering satellites spaced at 3° interval on each side of the "wanted" satellite.

If all the five satellites radiate with the same EIRPs toward the Earth station, at the same frequency, then expected level of interference with respect to the wanted signal can be calculated.

3.2.5.4.2 Co-Channel Interference

Co-channel interference occurs if external interference from other transmitters (e.g., terrestrial microwave transmitters) is at the **same frequency** as the signal of interest. Interference that is near the frequency of the signal is called **adjacent channel interference**. Co-channel interference may be caused by either of the following: harmonics from a different type of system unintentional radiators signals from a similar system that are some distance away (frequency reuse). In each case, the interference is received within the operating bandwidth of the receiver. Unlike thermal noise, co-channel interference cannot be reduced by simply increasing carrier power because increasing the carrier power at one nominal bandwidth increases the likelihood of interference with the carriers that are adjacent in frequency to the wanted carrier.

Satellites in geosynchronous orbits are co-located using orthogonal polarization to place more number of satellites to meet the demand. But while co-locating satellites operating in the same frequency band, careful system analysis and optimization is required to deal with co-channel interference. Since co-channel interference enters the receiver at or near the center of its bandwidth, the receiver filter does not attenuate it. Adjacent channel interference can be a problem depending upon the spectral properties of the receiver filter. Adjacent channel interference enters the receiver at a nearby frequency and therefore is attenuated by the receiver filter with a sharper roll-off. If the filter roll-offs are not sufficiently attenuated, the interference can cause undesired operation. Figure 3.62 shows receiver filter response with co-channel and adjacent interference.

Ideally, the power of a carrier transmitted by an Earth station is entirely contained within a fixed range of frequencies, or bandwidth (e.g. 6,000 MHz ± 500 kHz), and is zero outside of this range. This permits carriers from different Earth stations to be placed side by side in frequency with no interference between them, so long as their bandwidths do not overlap. However, in practice, some carrier power is radiated outside of the nominal bandwidth of the carrier and this can interfere with the carriers that are adjacent in frequency to the wanted carrier. The principal factor that governs the amount of interference that is generated in this way is the output back-off of the Earth station's HPA. The larger the back-off, the lower the potential for causing adjacent channel interference. Figure 3.63 shows co-channel interference in the desired satellite link due to the unwanted emissions from the uplink Earth station, (E/S2) of the other co-located satellite.

3.2.5.4.3 Cross-Polarization Interference

Most satellite systems employ opposite polarization states to make most efficient use of the frequencies that are available for transmission and reception. This means that two signals may share the same frequency within a satellite system, so long as they employ opposite polarization states (e.g., horizontal linear polarization and vertical linear polarization). Theoretically, each signal can be received without interference from the co-frequency signal on the opposite polarization. However, being practical devices, the Earth station and satellite antennas and their feeds are not able to perfectly separate the two polarization states, which results in a small proportion of the unwanted **"cross-polar"** signal being transmitted or received along with the wanted

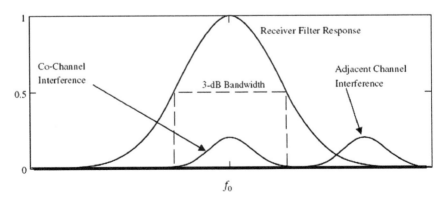

FIGURE 3.62
Receiver filter response with co-channel and adjacent channel interference.

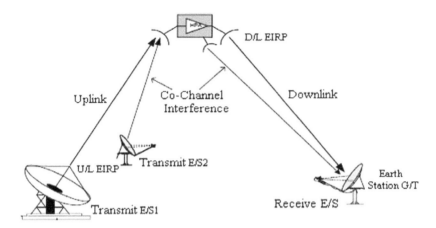

FIGURE 3.63
Co-channel interference from Earth station of co-located satellite.

signal, causing **cross-polarization interference**. Since the polarization performance of the satellite antennas is fixed, cross-polarization interference is maintained at an acceptable level by ensuring that the Earth station antenna has adequate cross-polar performance ("cross-polar discrimination, XPD"). If the XPD level of an uplink antenna is less than 30 dB, antenna transmits both vertical and horizontal polarizations. Polarization discrimination between co-polar and cross-polar signals is important in a dual polarization frequency reuse satellite communication system. Figure 3.64 illustrates polarization discrimination in an antenna radiation pattern.

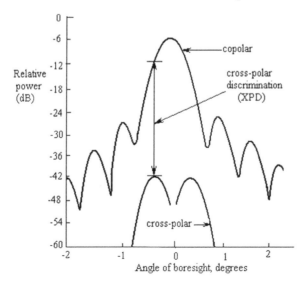

FIGURE 3.64
Polarization discrimination in an antenna radiation pattern.

In the case of linear polarization, it is also very important to ensure that the antenna feed is properly aligned with the linear polarization plane of the satellite antennas (by rotating the feed), otherwise the cross-polar discrimination of the antenna will be poor and this may lead to unacceptable levels of cross-polarization interference. The optimum isolation is received when both the transmitting (causing interference) and the receiving (victim) antennas have a similar cross-polarized response. A typical isolation requirement is about 27 dB for circular polarization and 33 dB for linear polarization. Low side-lobes and cross-polarization are necessary to prevent excessive interference among beams.

Interference between Two Neighboring Satellite Systems in GEO.

Two neighboring satellite systems in space may cause unacceptable levels of interference to each other. This may occur due to either of the following:

- Inadequate angular separation between the two transponders
- Bad antenna pointing
- Antenna beams not sufficiently concentrated in the direction of the wanted satellite

Uplink Interference: The satellite uplink comprises transmitting uplink station, and a satellite transponder which receives the signal. The uplink signal transmitted to the satellite transponder is called the "wanted signal" and the satellite transponder is called the "wanted satellite". There are, however, a number of other satellite transponders either co-located or in the vicinity of the wanted one, each of which is receiving its own signals. These unwanted signals may enter the receiver of the wanted satellite transponder causing interference to the wanted signal. This phenomenon is illustrated in Figure 3.65. Satellite B is supposed to receive a signal from Earth station 2, but it is also receiving an interfering signal from unwanted Earth station 1.

Downlink Interference: The satellite antenna cannot be pointing directly at all the Earth stations to which it is transmitting. There is a resultant angular difference between the boresight of the satellite antenna and the direction of the receiving Earth station given by O_3 in Figure 3.66.

3.2.5.4.3.1 The Overall Link Performance   The signal-to-noise plus interference ratio for the overall link is calculated by combining the separate uplink and downlink contributions. For interference-free transmission and reception, the signal-to-noise density ratio for the overall link may be expressed as

$$\left(\frac{C}{N_0} \right)_{Overall} = \left[\left(\frac{C}{N_0} \right)_{V/L}^{-1} + \left(\frac{C}{N_0} \right)_{D/L}^{-1} \right]^{-1}$$

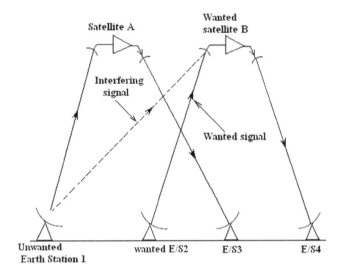

FIGURE 3.65
Uplink interference.

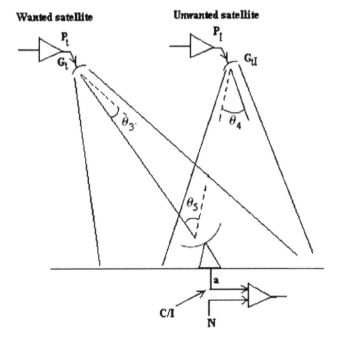

FIGURE 3.66
Composite view of downlink.

But for a system affected by interference, the overall signal-to-noise density plus interference ratio is expressed as

$$\left(\frac{C}{N_0 + I_0}\right)_{\text{Overall}} = \left[\left(\frac{C}{N_0 + I_0}\right)_{V/L}^{-1} + \left(\frac{C}{N_0 + I_0}\right)_{D/L}^{-1}\right]^{-1}$$

And in the bandwidth of interest, signal-to-noise plus interference ratio is given by

$$\left(\frac{C}{N + I}\right)_{\text{Overall}} = \left[\left(\frac{C}{N + I}\right)_{V/L}^{-1} + \left(\frac{C}{N + I}\right)_{V/L}^{-1}\right]^{-1} \text{dBHz}$$

Figure 3.67 illustrates combination of up and downlinks in presence of an interfering signal with power I W in bandwidth B Hz.

Overall Energy of Bit-to-Noise Density Ratio is a standard product over the sum relationship and is expressed mathematically as, where all ratios

$$\left(\frac{E_b}{N_0 + I_0}\right)_{\text{Overall}} = \frac{\left(\frac{E_b}{N_0 + I_0}\right)_{V/L} \times \left(\frac{E_b}{N_0 + I_0}\right)_{D/L}}{\left(\frac{E_b}{N_0 + I_0}\right)_{V/L} + \left(\frac{E_b}{N_0 + I_0}\right)_{D/L}}$$

$$\left(E_b / N_0 + I_0\right)$$

are in absolute values.

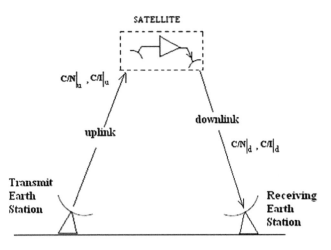

FIGURE 3.67
Combination of up and downlinks.

3.2.5.5 Broadband Access

Access to the American and European Internet backbone by foreign Internet service providers (ISPs) (also known as "broadband trunking") provides incremental opportunity for FSS operators. This application involves broadband trunking to regional hubs that extend spokes out to ISPs and extension links for ISPs seeking new territory. Satellites currently transmit a significant amount of Internet Protocol (IP) volume, especially in markets where fiber is not built-out. As mentioned earlier, satellites can provide a more cost-effective solution where terrestrial connectivity is unavailable due to the difficulty of service provision or cost issues; or serve as a supplement to existing terrestrial IP capacity. Direct links between an ISP point of presence and a major Internet backbone network can typically be achieved at data rates between 64 kbps and 45 Mbps, expanding on the amount of satellite capacity the ISP leases.

It is expected that broadband trunking will be a growth area for satellites as people in increasingly remote areas of the world begin to access the Internet. The UN information and communication technology (ICT) program intent was to provide Internet access to majority of the member countries by 2015. However, one difficulty with broadband trunking is the relatively small size of some of the customers, particularly foreign ISPs. Many of these companies cannot be relied on to commit to long-term contracts given that fiber alternatives may soon be available to some of them. To a certain extent, many of these customers must continuously be replaced by new customers who are in increasingly remote areas. Moreover, many contract for relatively small amounts of bandwidth, often fractions of a transponder.

Broadband connectivity via satellite-based consumer services has made significant headway in the last few years. Systems like Wildblue (United States), Direcway/Spaceway from Hughes Network Systems (HNS) (United States), Starband (United States), and IPStar (Southeast Asia) provide two-way broadband access direct to consumers. Over the long term, developing markets offer greater growth potential due to their large geographical areas and underdeveloped telecommunications infrastructure compared to rural areas. Strong demand exists in regions where terrestrial networks are inadequate and per-capita incomes do not justify infrastructure investment.

3.2.5.6 Very Small Aperture Terminal System

Many of the satellite services are provided through the Very Small Aperture Terminal system (VSAT) network to businesses, home users, and institutions. The driving force for such a trend could be summarized in the following:

The desire of governments to ensure that all their citizens are able to obtain equivalent ICT services regardless of their geographical locations. This is also driven by a desire to stop the urbanization process that is so destructive, especially in the developing countries.

The desire by large corporations for redundant and alternative networks, independent of the terrestrial infrastructure.

The need for a connectivity that is always available, regardless of geographical and political constraints.

The increased need for high-speed services connectivity and multimedia direct to the end-users.

The need for more connectivity.

VSAT networks provide corporations with the ability to create private global data networks in a relatively short period of time and at a low cost that is otherwise very difficult. Such networks are well suited for transmissions to multiple locations domestically, regionally, or even globally. VSATs are inherently optimized for broadcast applications, exploiting the point-to-multipoint advantages offered by satellite communications. Depending on the number of remote sites, VSAT networks could be designed either in Star Topology or Mesh Topology. VSATs are popular for point-to-point applications that are not bandwidth intensive, such as credit card authorizations. VSAT's star topology is typically cost-effective for networks of more than 200 sites, as the cost of the space segment is shared by a large number of sites. If the numbers of sites are less than 200, the mesh topology would be more cost-effective. In the star topology, as shown in Figure 3.68,[70] there is one, or if the network is very large more than one, hub station and hundreds or even thousands of remote stations that are all connected through the hub station. In the mesh network, as shown in Figure 3.69,[70] the remote sites have larger

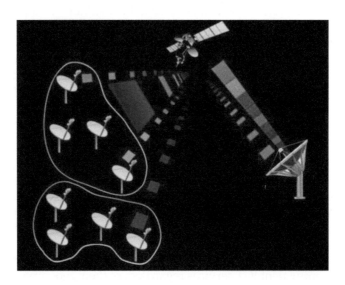

FIGURE 3.68
VSAT star network.

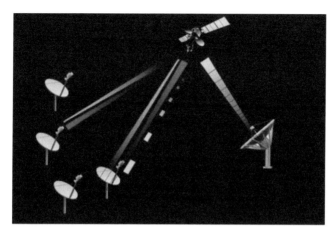

FIGURE 3.69
VSAT mesh network.

antennas with more software capabilities that make it possible for the remote stations to communicate with one another directly with no need of a central hub station. The main factor that determines which topology to use is the total cost of the network compared to star network.

This is how a star data, TDM/TDMA VSAT network works using a hub station, usually 6 m or more in diameter and small VSAT antennas (between 0.75 and 2.4 m). In this setup, all the channels are shared and the remote terminals are online.

TDM/TDMA systems have historically competed with terrestrial X.25 or frame relay connections, but as VSAT transmit data rates have risen to 2 Mbps or more and receive rates begin approaching 100 Mbps, DSL and MPLS services have become the main competitors in most markets.

The Mesh network as described above is shown in Figure 3.69.

Mesh networks that use capacity on a demand-assigned multiple access (DAMA) basis, use the master control station that acts as a controller and facilitator rather than a hub. Mesh connections take a little time to set up and are often compared with a terrestrial dial-up connection. Major advantages of VSAT networks could be summarized as follows:

- The "last-mile" solution used for their end-to-end service capability.
- Cost insensitive to distance and user density.
- Technology consistency meaning one set of protocols and hardware can be used globally with one service provider.

3.2.5.7 Satellite Multicasting

Multicasting allows ISPs, content providers, and corporations to use satellites to avoid terrestrial bottlenecks. Content is uplinked to satellites via Earth stations and then multicast out to ISPs, cable, and DSL points of presence at the periphery of the local loop. One of the benefits of this delivery mechanism is that it bypasses bottlenecks in the fiber backbone because the content is distributed closer to the end-user. Also, because content is broadcast to multiple points of presence, it is more economical than each user individually requesting the content and paying fiber backhaul fees. The economics is particularly compelling for applications such as video which is bandwidth intensive. Although multicasting systems such as Akamai Technologies (which operates a system of 15,000+ widely dispersed servers on which its customers deploy their Web content and applications in order to increase the performance and reliability of their websites), Cidera, and Net/36 did not live up to their early promises in the 1990s, interest in multicasting is slowly reemerging.

Intelsat is the largest satellite communication organization, worldwide, that provides a wide spectrum of FSSs. Figure 3.70[71] shows Intelsat's global coverage.

Intelsat provides services in the following categories:

Media Services:

Video distribution

Video contribution

Direct-To-Home (DTH)

High Definition (HD)

Intelsat's managed media services:

Hybrid satellite, fiber, and teleport-managed services

Part-time managed services

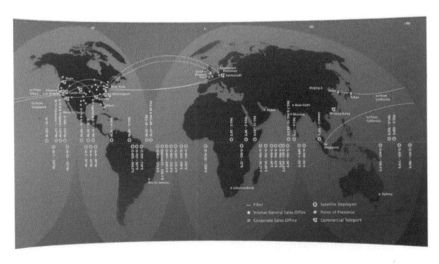

FIGURE 3.70
Intelsat global coverage.

Satellite-related Services: Intelsat's satellite-related services are:

Professional services

Commercial hosted payloads and instruments

Satellite procurement and oversight

Launch services procurement and oversight

Ground system solutions

Transfer orbit support services (TOSS)

In-orbit testing (IOT)

3.2.5.7.1 Satellite and Payload Operations

To get a sense of how Intelsat's coverage has grown to the present level of universality and becoming the largest satellite organization in the world, in the following Intelsat's organizational chart, a brief history, and generation of its satellites will be introduced. Intelsat's organizational chart is shown in Figure 3.71.[72]

Intelsat's milestones since its inception could be summarized as follows:

The 1960s

In 1965, the organization launched the world's first commercial communications satellite, the "Early Bird." Within 4 years, the organization established the first global satellite communications

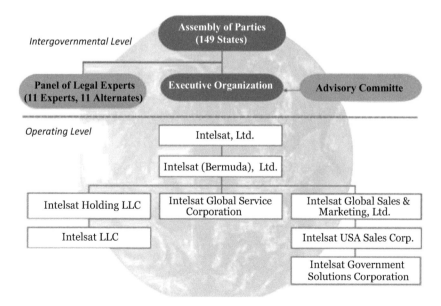

FIGURE 3.71
Intelsat's organizational chart.

system, with satellites covering the three major ocean regions. In 1969, the organization transmitted footage of the historic Apollo XI lunar landing.

The 1970s

In 1974, the organization created the world's first international digital voice communications service. Later that year, the organization activated a direct "hot line" link between the White House and the Kremlin. In 1978, an estimated 1 billion people in 42 countries watched World Cup soccer matches via the organization's satellite system.

The 1980s

In the 1980s, the organization introduced more powerful satellites that allowed broadcasters to use small, easily transportable Earth stations to broadcast major news events, live, from anywhere in the world for the first time. In 1987, the organization engineered the largest international teleconference in history, linking over 50,000 people in 79 cities to discuss world hunger.

The 1990s

In the early 1990s, the organization's satellites were used for commercial Internet applications and multimedia. In 1997, the organization offered the world's first, global, pay-as-you-go, shared-access satellite service for low-demand areas. This enabled public telecom operators to connect many rural villages and remote areas for the first time.

2000 and Beyond

In the context of the World Summit on the Information Society (WSIS) (Geneva 2003–Tunis 2005), the Global Broadband Satellite Infrastructure (GBSI) Initiative highlights the role of satellites in ensuring the provision of high-speed Internet access to allow the dissemination of ICT applications to everyone and everywhere.

Intelsat's generations of satellite communications are listed in Table 3.18[73] at the end of this chapter. In the following section, a brief description of the

TABLE 3.18

Example of Atmospheric Attenuation

Atmospheric Loss (dB)	Frequency (f) in GHz
0.25	$2 < f < 5$
0.33	$5 < f < 10$
0.53	$10 < f < 13$
0.73	$13 < f$

Earth station design is discussed. Having a good understanding about the parameters involved in satellite link analysis will help us to better understand how a satellite communication system functions and what are the limiting factors in the design of Earth station subsystems. Intelsat as the largest satellite communication organization worldwide with over 50 satellites at the time of this writing uses such material in training their engineers and technicians to design and maintain satellite Earth stations. There are numerous link analysis softwares online that interested readers can use to get a better feeling of how these parameters would impact the satellite link and the quality of services that it provides.

3.2.5.8 Earth Station Design

3.2.5.8.1 Introduction

A satellite link is defined as an Earth station–satellite–Earth station connection. The Earth station–satellite segment is called the uplink and the satellite–Earth station segment is called the downlink. The Earth station design consists of the Transmission Link Design, or Link Budget, and the Transmission System Design. The satellite Link Budget establishes the resources needed for a given service to achieve the performance objectives. The Transmission System Design, however, establishes the equipment characteristics necessary to meet the performance objectives for the services to be provided, such as the HPA-rated power and the low noise amplifier (LNA) temperature. During the analysis, trade-offs can be made to achieve a balance between cost and performance.

In this part of the book, we take a look at a section of a handbook that has been prepared under Intelsat's Assistance and Development Program (IADP) which is used as a reference handbook for courses on Earth station communications technology.[74]

- *Performance Objectives*
  Performance objectives for digital links consist of the following factors:

  Bit error rate (BER) for normal operating conditions

  Link availability which translates into the percentage of time that the link has a BER better than a specified threshold level

- *Link Budget*
  The satellite link illustrated in Figure 3.72 is composed primarily of three segments: the transmitting Earth station and the uplink media, the satellite, and the downlink media and the receiving Earth station. The overall carrier level received at the end of the link is simply the addition of the losses and gains in the path between transmitting and receiving Earth stations, most of which are shown in Figure 3.72.

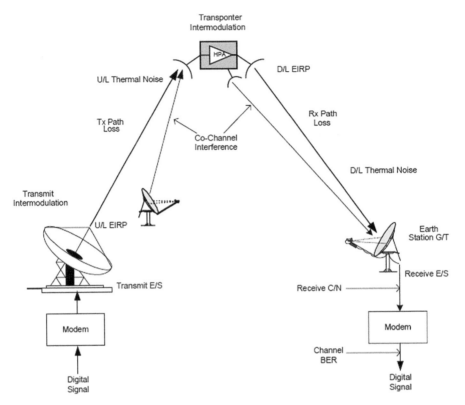

FIGURE 3.72
Typical satellite link. (Intelsat Earth Station Technology, Revision 5, June 1999.[74])

- *Carrier-to-Noise Relationship*

  The transmission performance of the RF portion of the satellite system is established by the basic carrier-to-noise relationship in a system and is defined by the receive carrier power level compared to the noise at the receiver input. Figure 3.73 illustrates how link components affect the receive carrier-to-noise ratio, and finally, the service quality. For example, the downlink thermal carrier-to-noise ratio is

$$C/N = C - 10\log(\text{kTB}) \tag{3.3}$$

where

    C = Receive power in dBW

    K = Boltzmann constant, 1.38E-23 W/K/Hz

    B = Noise bandwidth (or occupied bandwidth) in Hz

    T = Absolute temperature of the receiving system in °K

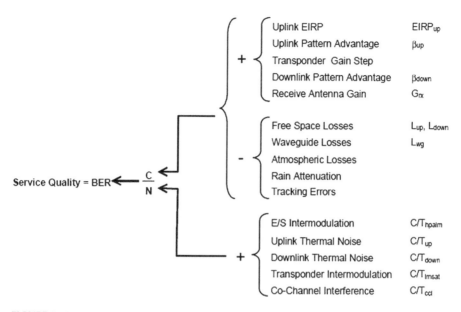

FIGURE 3.73
Link parameters' impact on service quality. (Intelsat Earth Station Technology, Revision 5, June 1999.[75])

3.2.5.8.2 Link Equation

In its general form, the link equation can be expressed as follows:

$$C/N = EIRP - L + G - 10\log kTB \qquad (3.4)$$

where

EIRP = Equivalent isotropically radiated power (dBW)

L = Transmission losses (dB)

G = Gain of the receive antenna (dB)

The first three terms contribute to the received carrier power, and the final term is the noise power of the receiving system. It is to be noted that the link equation applies to both uplink and downlink.

Another point to note is that the transmission losses, L, are defined as the sum of the free space transmission loss plus any additional path losses that would include such factors as the loss due to rain.

- *Equivalent Isotropically Radiated Power (EIRP)*
  The EIRP as a function of the antenna transmits gain, G_T and the transmitted power P_t, fed to the antenna is expressed as

$$EIRP(dBW) = 10\log P_T(dBW) + G_T(dBi) \qquad (3.5)$$

where

P_T (dBW) = Antenna input power in dBW

G_T (dBi) = Transmit antenna gain in dBi

Because an excessive EIRP will cause interference to adjacent and co-channel carriers, while a low EIRP will result in poor quality performance of the service, the EIRP must be accurately controlled.

- *Antenna Gain*
  Antenna gain, referred to as an isotropic radiator, is defined by

$$G(\text{dBi}) = 10\log\eta + 20\log f + 20\log d + 20.4\,\text{dB} \qquad (3.6)$$

where

η = Antenna efficiency (typical values are 0.55–0.75)

d = Antenna diameter in m

f = Operating frequency in GHz

- *Transmission Losses*
  Transmission losses generally consist of the following four components:

$$L = L_o + L_{\text{atm}} + L_{\text{rain}} + L_{\text{track}} \qquad (3.7)$$

where

L_o = Free-space loss

L_{atm} = Atmospheric losses

L_{rain} = Attenuation due to rain effects

L_{track} = Losses due to antenna tracking errors

- *Free-Space Loss*
  In an isotropically radiating antenna, the beam power, PT will spread as a sphere in which the antenna is the center. The power at a distance D from the transmission point is, therefore, given by

$$W = P_T / 4\pi D^2 \quad \left(\text{W/m}^2\right) \qquad (3.8)$$

As the transmit antenna focuses the energy, the equation will change to

$$W = G_T P_T / 4\pi D^2 \quad \left(\text{W/m}^2\right) \qquad (3.9)$$

or

$$W_{\text{dBW/m}^2} = \text{EIRP}_{\text{dBW}} - 20\log D - 71\,\text{dB} \qquad (3.10)$$

where

GTPT = EIRP

W = Illumination level

D = Distance in km

71 dB = 10 log ($4\pi * 106$)

Because the amount of collected signal depends on the receiver antenna size, the received power PR is expressed as

$$P_R = W * A_e \qquad (3.11)$$

where A_e = effective aperture of the receive antenna = $(\lambda^2/4\pi)/GR$.
    Then,

$$P_R = \left[G_T P_T / 4\pi D^2 \right] \left[(\lambda^2/4\pi)/G_R \right] \qquad (3.12)$$

$$P_R = G_T P_T \left(\lambda/4\pi D \right)^2 G_R \qquad (3.13)$$

The expression $[4\pi D/\lambda]^2$ is known as the basic free-space loss L_o which is expressed in decibels as

$$L_o = 20\log D + 20\log f + 92.5\,\text{dB} \qquad (3.14)$$

where

D = Distance in km between transmitter and receiver, or slant range

f = Frequency in GHz

92.5 dB = 20 log [$4\pi * 109 * 103/C$]
Expressing Equation 3.13 in decibels, we have

$$P_{R\ \text{dBW}} = \text{EIRP} - L_o + G_R \qquad (3.15)$$

In Equation 3.15, if GR was the gain for a $1\,\text{m}^2$ antenna with 100% efficiency, PR will then become the illumination level per unit area in dBW/m². The illumination level in Equation 3.10 can also be expressed as

$$W_{\text{dBW}/\text{m}^2} = \text{EIRP} - L_o + G_{1\,\text{m}^2} \qquad (3.16)$$

- *Atmospheric Losses*
  Signal losses can also occur through absorption by atmospheric gases such as oxygen and water vapor.
    This characteristic of the signal depends on the frequency, elevation angle, altitude above sea level, and absolute humidity. The effect

of atmospheric absorption is negligible at frequencies below 10 GHz, and its importance increases with frequencies above 10 GHz, especially for low-elevation angles. Table 3.18 shows an example of the mean value of atmospheric losses for a 10° elevation angle.

- *Rain Effects*
  Rainfall has an important climatic effect on a satellite link. Rain results in attenuation of radio waves by scattering and by absorption of energy from the wave. Rain attenuation increases with the frequency and it gets worse as we go from C-band to K_u-band and further to K_a. In order to overcome the additional attenuation induced by rain, enough extra power must be transmitted to provide adequate link availability. Knowing that the prediction of rain attenuation is a statistical process, many models have been developed confirming experimental observation. Parameters considered in these models relate to operating frequency, rain rate statistics by geographic location, as well as the proposed link availability. To realistically determine the link availability and establish the appropriate link margin, a reliable prediction of attenuation by rain is desirable.

- *Tracking Losses*
  In an established satellite link, the ideal situation is to have the Earth station antenna aligned for maximum gain. In such links, there is a small degree of misalignment that causes the gain to drop by a few tenths of a dB. This reduction of gain can be estimated from the antenna size, the tracking type, and accuracy. This loss would affect both the uplink and downlink calculations. Typical values can be found in Tables 3.19 and 3.20 for C-band and K_u-band antennas. Larger antenna diameters will always require tracking, and such misalignment losses can be set as 0.5 dB for both uplinks and downlinks.

- *Pattern Advantage*
  The satellite antenna pattern has a defined beam edge with referenced values of EIRP, G/T, and flux density (see Figure 3.74). In the

TABLE 3.19

Earth Station Performance Characteristic, C-Band (C-Band, Antenna Efficiency 70%)

Antenna Diameter (m)	Tx Gain 6 GHz (dBi)	Rx Gain 4 GHz (dBi)	Uplink Losses (dB)	Downlink Losses (dB)	Tracking
1.2	35.6	32.1	0	0	Fixed
1.8	39.2	35.6	0	0	Fixed
2.4	41.7	38.1	0.4	0.2	Fixed
3.6	45.6	42.1	0.7	0.4	Fixed
7	51	47.4	0.9	0.9	Manual[a]
11	54.9	51.4	0.5	0.5	Step track

[a] Manual tracking requires weekly E-W angle adjustments.

TABLE 3.20

Earth Station Performance Characteristic, K_u-Band (K_u-Band, Antenna Efficiency 60%

Antenna Diameter (m)	Tx Gain 14 GHz (dBi)	Rx Gain 12 GHz (dBi)	Uplink Losses (dB)	Downlink Losses (dB)	Tracking
1.2	42.6	40.5	0.4	0.2	Fixed
1.8	46.1	44	0.7	0.5	Fixed
2.4	48.2	46.6	1.1	0.8	Fixed
3.7	52.5	50.3	1.2	0.9	Manual[a]
5.6	56.1	53.9	0.8	0.7	Manual[a]
7	58	55.8	0.5	0.5	Step track
8	59.2	57	0.5	0.5	Step track

[a] Manual tracking requires weekly E-W angle adjustments.

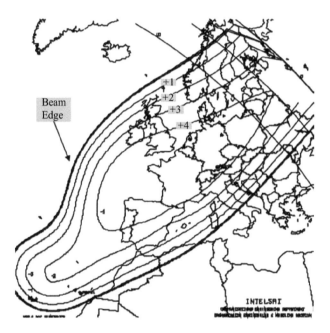

FIGURE 3.74
Example of pattern advantage. (Intelsat Earth Station Technology, Revision 5, June 1999.[75])

satellite link analysis, adjustment factors are applied to account for the location of an Earth station within the satellite beam. These adjustment factors, which are referred to as beta-factors, aspect correction, or pattern advantage, apply to all satellite beams, with the exception of the global beams. By definition, the beta-factor is the difference between the satellite beam edge gain and the gain in the direction of an Earth station.

- *System Noise Temperature*
  The system noise temperature of an Earth station consists of the following components:

  Receiver noise temperature

  The noise temperature of the antenna, which includes the feed and waveguides

  The sky noise picked up by the antenna

  These elements could be expressed in the following equation:

$$T_{system} = T_{ant}/L + (1 - 1/L)T_o + T_e \tag{3.17}$$

  where

  L = Feed loss in numerical value

  T_e = Receiver equivalent noise temperature

  T_o = Standard temperature of 290°K

  T_{ant} = Antenna equivalent noise temperature as provided by the manufacturer.

- *Antenna Noise Temperature*
  The noise power into the receiver (i.e., the LNA) due to the antenna is equivalent to that produced by a matched resistor at the LNA input at a physical temperature of T_{ant}.

  The antenna noise temperature is a complex function of several parameters such as antenna gain pattern, background noise, temperature of the sky, equivalent atmospheric noise temperature, and noise temperature of the Sun. Figure 3.75 shows a typical curve variation of the antenna noise temperature with the antenna elevation angle. It is usually a minimum at zenith, typically 15–20°K for a low loss antenna with low side-lobes and increases noticeably as the elevation angle falls below 10°.

- *G/T or "Figure of Merit"*
  The G/TdBK which is commonly known as the Figure of Merit is also referred to as the "goodness" measurement of a receiver system. Intelsat requires a specific G/T for all of its standard Earth stations that are used globally and as presented in the/ESS documents. G/T is expressed in dB relative to 1°K. One should note that the same system of reference point, such as the receiver input, for both the gain and noise temperature must be used.

  G/T could be expressed in terms of the receiver gain and system noise temperature as

$$G/T = G_{tx} - 10LogT_{sys} \tag{3.18}$$

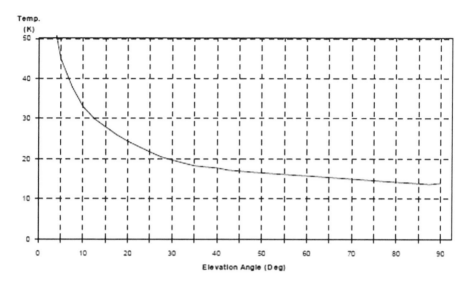

FIGURE 3.75
Noise temperature of an antenna as a function of elevation angle.

where

G_{tx} = Receive gain in dB

T_{sys} = System noise temperature in °K.

Because the antenna gain is frequency dependent, the G/T must be normalized to a known frequency (normally either 4 or 11 GHz which are the downlink frequencies for the C-band and K_u-band) by subtracting from Equation 3.18 the factor $20 \log f/f_0$ (f_0 being 4 or 11) where f is the frequency in GHz.

- C/N—*Carrier-to-Noise Ratios*
  C/N could be expressed in the following form when the kTB product is unfolded under the logarithm:

$$C/N = EIRP - L + G - 10\log T - 10\log B \qquad (3.19)$$

As was mentioned above the $G - 10\log T$, is the figure of merit, for example,

$$C/N = EIRP - L + G/T - 10\log K - 10\log B \qquad (3.20)$$

where

L = Transmission losses

G/T = Figure of merit of the receiver

K = Boltzmann constant

B = Carrier occupied bandwidth

Because B, the receiver bandwidth is often dependent on the modulation format, we can isolate the link power parameters by normalizing out the bandwidth dependence resulting in the new relation which is known as carrier-to-noise density ratio (C/N_0).

$$C/N_0 = EIRP - L + G/T - 10\log K \qquad (3.21)$$

Note that

$$C/N = C/T - 10\log kB \qquad (3.22)$$

Expressing C/T as a function of C/N, and replacing C/N with the right side of the link equation, results:

$$C/T = EIRP - L + G/T \qquad (3.23)$$

C/T indicates the level of carrier power required for a given G/T. For example, the C/T for an acceptable TV signal might be $-140\,dBW/°K$ as compared to $-150\,dBW/°K$ for a digital carrier, although the C/N for the two carriers at the input of the demodulation may be the same. C/N is therefore the characteristic for each carrier size and type. The ratio C/N_0 also allows us to compute the receiver bit energy-to-noise density ratio—that is, E_b/N_0 which could be expressed as

$$E_b/N_0 = C/N_0 - 10\log(\text{digital rate}) \qquad (3.24)$$

The term "digital rate" is used here because E_b/N_0 can refer to different points with different rates in the same modem.

- *Impact of G/T on Service Economics*
  Because higher satellite EIRP implies higher operation costs, one can conclude that in Equation 3.23, a higher C/T, which is expressed in terms of EIRP, also results in an increase in cost. On the other hand, the G/T represents the capital expenditure, because higher G/T means larger antenna or better LNA, which is reflected in the cost of the equipment. For long-term usage of an Earth station facility, it may be more economical to build a larger antenna that will require a lower downlink EIRP, compared to a smaller antenna that will require higher satellite EIRP for the same quality of service. Note that in some cases the Earth station G/T could be improved by using a better LNA. For example, an Earth station with a receive gain of 53 dBi, antenna noise of 25°K at C-band, feeder noise temperature of 5°K and LNA temperature of 80°K would have

$$G/T = G_{ant} - 10\log(T_{ant} + T_{feed} + T_{LNA}) \qquad (3.25)$$

$$G/T = 53 - 10\log(25 + 5 + 80) = 32.6\,dB/°K$$

This antenna would be classified as an Intelsat standard B antenna. To see the impact of a better LNA, we can remove the LNA and replace it with a 30°K LNA. The resulting G/T could now be written as:

$$G/T = 53 - 10\log(25 + 5 + 30) = 35.2\,\text{dB/°K}$$

This reclassifies the antenna as an Intelsat standard A.

- *The Satellite Transponder*
  The function of satellite transponders is the same as that of a radio relay repeater—that is, receive transmission from the Earth and retransmit to the Earth after frequency translation and amplification. Satellite resources are shared among many Earth stations, with different categories of standard A, B, C, D, E, and F, and therefore, with different satellite requirements. Other than the bandwidth, the parameters for a given transponder are

  1. Saturation flux density (SFD, dBW/m²)
  2. Receive G/T (dB/°K)
  3. Saturation EIRP (dBW)

  SFD is the total power flux density arriving at the satellite from the Earth segment that will produce the saturation EIRP from the satellite.

- *Transponder Operating Point*
  Because the transponder output power amplifier is not a linear device, it must be operated below the saturation point to avoid nonlinear distortions. To achieve the operating point an input and output backoff (IBO and OBO, respectively) will be required. This is shown in Figure 3.75. This is an unavoidable waste of available power in a typical TWTA.
  The input backoff is defined as the ratio of SFD to the operation flux density for a given carrier. The output backoff is defined as the ratio of saturation EIRP to the operation EIRP for a given carrier. In terms of the input backoff, the output backoff is defined as:

  $$\text{OBO} = \text{IBO} - \text{X} \tag{3.26}$$

  "X" is the gain compression ratio between the IBO and OBO and its value is different for single carrier and multicarrier operation as shown in Figure 3.76.

- *Transponder Operating EIRP*
  The operating satellite EIRP (EIRPop) is calculated from Equation 3.26 as

  $$\text{EIRP}_{\text{op}} = \text{EIRP}_{\text{saturation}} - \text{OBO} \tag{3.27}$$

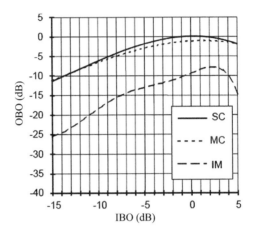

FIGURE 3.76
Transponder transfer characteristic and IM products generation.

- *Noise Components of the Link Budget*
  The noise components of the link budget are made up of the following.
- *Uplink Thermal Noise*
  Uplink thermal noise is caused by the inherent noise in the satellite receiving system.
      The antenna pointing errors and rain attenuation should be taken into account when calculating uplink C/T. Typically, 0.5–1.0 dB is left for pointing errors. As was explained before, the signal attenuation increases with frequency, so in the C-band, although it is normal to ignore rain attenuation, but in areas with very high rainfall rates, the effect should be considered. In the K_u-band, margins of 2–4 dB are normal for attenuation due to rain.
      Uplink thermal noise is calculated using the following formula:

$$C/T_{up} = EIRP_{up} - L_{up} + G/T_{sat} + \beta_{up} - m_{up} \qquad (3.28)$$

where

    $EIRP_{up}$ = Uplink EIRP
    L_{up} = Path loss for the uplink
    β_{up} = Uplink pattern advantage
    m_{up} = Margin for rain and tracking error, and so forth

Earth Station HPA Intermodulation Products

When wideband HPAs operate under multicarrier conditions intermediation products could be generated over the entire satellite frequency band (500 MHz). It should be noted that even if the user's plan involves only one carrier per HPA, there is still a potential for interference from intermodulation products generated from other stations operating under multicarrier conditions in the same uplink beam.

The C/THPAIM is derived from the HPA-IM limits:

$$C/T_{HPAIM} = EIRP - A + X - 192.6 \text{ dBW}/°K \tag{3.29}$$

where

A = HPA IM limit at 10° elevation angle

X = Correction factor for elevation angle and Earth station location:

$$X = 0.02(\alpha_u - 10) + \beta_{u+\gamma}\left[0.02(\alpha_d - 10) + \beta_d\right] \tag{3.30}$$

where

α_u = Elevation angle of the transmit E/S

β_u = Difference between the satellite receive beam edge of coverage gain and the gain in the direction of the transmit E/S, in dB

γ = The fraction of the downlink factor to be used in total factor adjustment

α_d = Worst located receiving E/S elevation angle

β_d = Difference between the satellite transmit beam edge of coverage gain and the gain in the direction of the worst located receiving E/S in dB

Co-Channel Interference

Co-channel interference is caused by carriers on the same satellite and at the same frequency, but in different up and down beams (frequency reuse), which are separated either spatially or by using the opposite polarization. This value is given as a carrier-to-interference (C/I) ratio in dB. To convert C/I to C/T, we can use the following equation:

$$C/T_{cci} = C/I + 10\log(OccBw) - 10\log k \tag{3.31}$$

where OccBw = Carrier-occupied Bw, and refers to the carrier for which the calculation is being performed.

Transponder Intermodulation

Transponder intermodulation is specified as a limit of EIRP density transmitted from the transponder in a 4 kHz bandwidth at the beam edge:

$$C/T_{\text{sat im}} = \text{EIRP}_{\text{down}} - \text{SAT}_{\text{im}} + \beta_{\text{down}} - 192.6\,(\text{dBW}/^\circ\text{K}) \qquad (3.32)$$

where

$\text{EIRP}_{\text{down}} = \text{Downlink EIRP}$

$\text{SAT}_{\text{im}} = \text{Specified intermodulation limit in 4 kHz}$

$\beta_{\text{down}} = D/L \text{ pattern advantage}$

- *Downlink Thermal Noise*
  As in the uplink thermal noise case, a margin should be considered to allow for rain attenuation and tracking errors. This noise is caused by the Earth station receiving system and can be expressed by the following equation:

$$C/T_{\text{down}} = \text{EIRP}_{\text{down}} - L_{\text{down}} + G/T_{E/S} - m_{\text{up}} \qquad (3.33)$$

where

$C/T_{\text{down}} = \text{Downlink EIRP}$

$L_{\text{down}} = \text{Downlink path loss}$

$G/T_{E/S} = \text{Earth station figure of merit}$

$m_{\text{up}} = \text{downlink margin for tracking and rain}$

Total Link Carrier-to-System Noise Temperature (C/T_T)

Summing up the above ratios will result in the total C/T_T for the whole link, using the equation:

$$1/(C/T_T) = 1/\left(C/T_{\text{up}}\right) + 1/\left(C/T_{\text{down}}\right) + 1(C/T_{\text{im } e/s})$$
$$+ 1(C/T_{\text{sat im}}) + 1\left(C/T_{\text{cci}}\right) \qquad (3.34)$$

Figure 3.77 shows these elements and the total C/T ratio, which is lower than the lowest C/T ratio, as the noise is additive.

A low EIRP level means low C/N_0, but as shown in Figure 3.77, a higher EIRP level does not necessarily mean better C/N_0.

3.2.5.8.2.1 High Power Amplifier (HPA) Sizing When deciding the required HPA size, the total EIRP for all carriers and the required backoff must be taken into account.

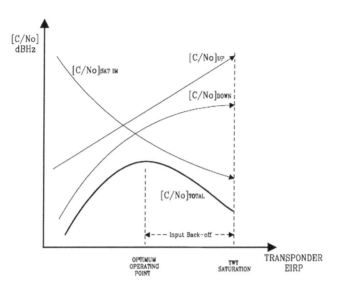

FIGURE 3.77
Variation of (C/N_0) as a function of carrier power.

In order to provide a better picture of these parameters, we look at the following example. Assuming that the HPA transmits two carriers having EIRP1 and EIRP2 levels, the total EIRP is calculated by converting the two carriers' powers to Watts. After calculating the total power required at the antenna input, the feed losses and output backoff of the HPA must all be taken into account as shown in Table 3.21.

In this section of the book, which was entirely allocated to the Earth station design and link budget calculations and analysis, we learned that link analysis equations are essential to analyze the system performance.

TABLE 3.21

High Power Amplifier (HPA) Sizing

Parameter	Formula	Value Unit
EIRP$_1$, carrier 1		60.1 dBw
EIRP$_2$, carrier 2	$= 10 \times \log (10 \, (\text{EIRP}_1/10) + 10 \times (\text{EIRP}_2/10))$	63.2 dBw
EIRP total through HPA (EIRP$_1$)		64.9 dBw
Antenna gain (G_{ant})	$= \text{EIRP}_1 - G_{ant} + L_f$	52.9 dB
Feed losses (L_f)		1.0 dB
Power required at HPA output (P_{req})		13.0 dBw
HPA back off (PA$_{OBO}$)	$= P_{req} + \text{PA}_{OBO}$	8.0 dB
Saturated HPA output power (P_s)	$= 10 \times (P_s/10)$	21.0 dBw
HPA size		126.8 w

The same information can be used in several other areas such as antenna acceptance tests, equipment requirements, satellite resources requirement calculations, network design, and cost estimates.

There are many softwares available online that could be used for the link budget calculations and analysis which will result in a better understanding of the impact of individual parameters on the overall performance of a satellite link. It is essential to understand the role of every single parameter in the overall satellite link analysis and their combined affects as well.

3.2.5.9 Satellite Services

3.2.5.9.1 Fixed Satellite Service

The allocation of radio frequencies is provided according to Article 5 of the ITU RR (edition 2012).

In order to improve harmonization in spectrum utilization, the majority of service-allocations stipulated in this document were incorporated in national Tables of Frequency Allocations and Utilizations which is with-in the responsibility of the appropriate national administration. The allocation might be primary, secondary, exclusive, and shared.

- primary allocation: is indicated by writing in capital letters (see example below)
- secondary allocation: is indicated by small letters
- exclusive or shared utilization: is within the responsibility of administrations

However, military usage, in bands where there is civil usage, will be in accordance with the ITU RR.

Satellites providing FSSs transmit radio communications between ground Earth stations at fixed locations. Satellite-transmitted information is carried in the form of RF signals. Any number of satellites may be used to link these stations. Earth stations that are part of FSSs networks also use satellite news gathering vehicles to broadcast from media events, such as sporting events or news conferences. In addition, FSS satellites provide a wide variety of services including paging networks and point-of-sale support, such as credit card transactions and inventory control.

Three main applications of FSS that have high potential of growth in the coming years are VSAT, broadband access, and satellite multicasting. In the following, these three applications are further explained.

3.2.5.9.2 Mobile Satellite Service (MSS)

According to the ITU RR, MSS is defined as a radio-communication service that is:

Between mobile earth stations and one or more space stations, or between space stations used by this service.

Between mobile Earth stations by means of one or more space stations. This service may also include feeder links necessary for its operation.

MSS uses portable terrestrial terminals that may be mounted on a ship, an airplane, or an automobile. MSS terminals may even be carried by an individual.

MSSs are a crucial component of the satellite communication industry and have proven efficacy for providing communication services across the globe, especially in servicing those areas with low-quality telecommunication infrastructure or limited communication options. Some of the main characteristics of MSS are:

- Very high bit rates,
- Channels dynamically are assigned to users, who are located in different locations or areas of the world,
- Insensitivity to distance makes the system a cost-effective service

Figure 3.78 below shows the global market for MSS from 2016 to 2026.

Among the MSS providers, Inmarsat has been the largest mobile satellite service provider for over 31 years and is recognized as the leader in this field. Inmarsat's fleet of 11 satellites provides seamless mobile voice and data communications around the world, enabling users to make phone calls or connect to the Internet, whenever and wherever they need, on land, at sea, or in

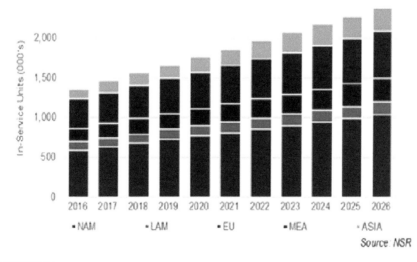

FIGURE 3.78
The global market for MSS.

the air. Inmarsat's Broadband Global Area Network service makes it possible for TV broadcasters to beam live breaking news from remote locations into millions of homes.

When a mobile phone call is made or a text message is sent from an aircraft, it is most likely that Inmarsat satellites are used to route the signal to the ground. Inmarsat, an international satellite communication organization similar to Intelsat, provides three types of services: land, mobile, and aeronautical. In the land services, sectors such as Oil and Gas, Media, Utilities, Mining, Aid, and Transportations use Inmarsat services. In these sectors high and low speed data and voice services are provided. In the maritime, services are provided to merchants, government, and used for fishing and leisure. Fleet Broadband and Fleet Phone are also provided by Inmarsat. In the aeronautical category of services, Inmarsat provides services to Air Transport, Business, and Government. Figure 3.79 shows the global coverage of Inmarsat Satellites.

3.2.5.9.3 *Broadcasting Satellite Service*

The broadcasting satellite service is another radio communication service in which signals transmitted by satellites are intended for direct reception by the general public using TVRO terminals. The satellites used for broadcasting are often called Direct Broadcast Satellites (DBS) that include individual DTH and Community Antenna Television (CATV). The new generation of

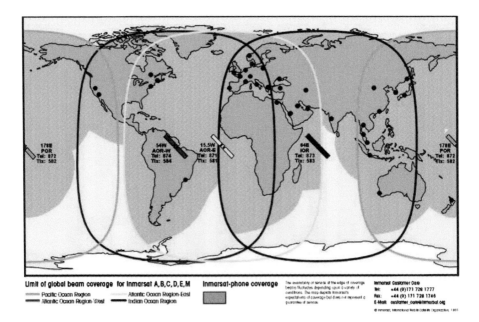

FIGURE 3.79
Inmarsat global coverage. (www.navtec.de/english/inmarsat.htm.[75])

BSS has a return link via satellite. Because the footprint of the satellite could be very large, the BSS can provide a worldwide coverage. The broadcast services include audio, television, and Internet services.

3.2.5.9.3.1 Orbital Spacing, EIRP, and Frequency Bands Because BSS are in nature high-power satellites, the orbital spacing is between $6°$ and $9°$ compared to $1°$ and $2°$ for FSS. This spacing is used to prevent adjacent satellite interference.

The EIRP for broadcasting satellite services are in the range of 51–60 dBW which is much higher than that of FSS. The frequency bands used for BSS are usually K_u band. Lower frequency band (i.e., C-band) requires larger size antenna which makes it difficult for installations. Also the interference with terrestrial systems is higher at lower frequencies because most of the terrestrial systems are designed at lower microwave frequencies. If we look at the other side of the spectrum, one might ask the question of why not use K_a-band for BSS because it provides much higher bandwidth. The answer is the fact that at higher frequencies, signals suffer more from attention, especially due to rain. For DBS services operating at K_a-band, the chances of losing the signal or receiving a very weak signal is higher during severe rainstorms. In order to benefit from the "frequency reuse" option, right-hand circular polarization and left-hand circular polarization are used.

In order to utilize the transponder capacity of satellites more efficiently, audio and video components of the television program are digitized to allow the signal compression be applied which greatly reduces the bandwidth required.

In a typical DBS system, the TV signal is received by a low noise block (LNB) converter, which is mounted at the focus of an antenna. The LNB downconverts the signal broadcast by the satellite to an intermediate frequency (IF), which in turn, is downconverted by a set-top converter to a signal that can be decoded into picture and sound by a television set. The DBS LNB converter is shown in Figure 3.80.

The LNB consists of four components: LNA, mixer, oscillator, and IF amplifier.

As can be seen in Figure 3.80, the LNA is the "front end" of the block downconverter that receives the broadcast television signal from the antenna, and

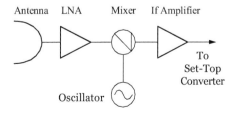

FIGURE 3.80
DBS low-noise block converter.

it amplifies the received signal sufficiently for further processing, with minimal signal-to-noise ratio degradation.

The noise figure and associated gain are the most important parameters of the LNB. The noise figure sets the sensitivity of the receiver and determines how small of an antenna can be used with the LNB. It is desired to lower the noise figure as much as possible for better performance.

The gain of the LNB on the other hand ensures that sufficient power will be available to drive the set-top converter. The following cascade noise equation demonstrates how the noise figure and associated gains work together.

$$NF = NF1 + \left[(NF2-1)/G1\right] + \left[\left[(NF3-1)/G1G2\right]\right] + \cdots \qquad (3.35)$$

Figure 3.81 shows a typical DBS receiving system.

A general configuration of a direct broadcast system is shown in Figure 3.82.

DBS systems use a geostationary satellite to receive television signals sent up from the Earth's surface. The received signal is amplified and transmitted back down to the receiving system which could be a community receiving system or simply an individual user. The satellite also down-converts the received signal frequency, 17.3–17.8 GHz uplink band, (in North America) to 12.2–12.7 GHz downlink band that is transmitted back down and is then picked up by the receiving antenna located atop an individual home or office. These antennas are usually in the form of a parabolic dish, shown in Figure 3.83, but flat square phased-array antennas are sometimes used.

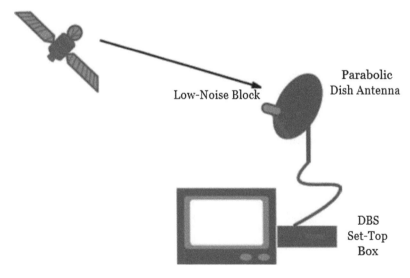

Parabolic
Dish Antenna

Low-Noise Block

DBS
Set-Top
Box

FIGURE 3.81
A typical DBS receiving system.

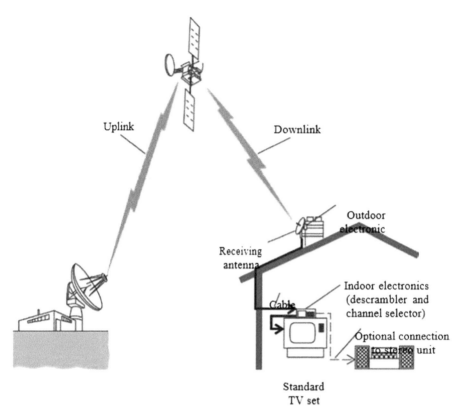

FIGURE 3.82
Typical DBS system configuration. (The Free Dictionary, Direct Broadcasting Satellite Systems, http://encyclopedia2.thefreedictionary.com/Direct+broadcasting+satellite+systems.[76])

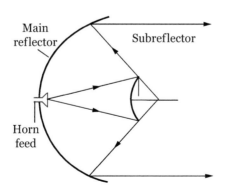

FIGURE 3.83
Cassegrain system.

TABLE 3.22

Broadcasting Satellite System (BSS) Frequency Bands

Region 1	11.7–12.5 GHz	Space-to-Earth
	14.5–14.8 GHz	Earth-to-Space
	17.3–18.1 GHz	Earth-to-Space
Region 2	12.2–12.7 GHz	Space-to-Earth
	17.3–17.8 GHz	Earth-to-Space
Region 3	11.7–12.2 GHz	Space-to-Earth
	14.5–14.8 GHz	Earth-to-Space
	17.3–18.1 GHz	Earth-to-Space

The receiving antenna is permanently pointed at the satellite which is in a geostationary orbit.

Table 3.22 shows the frequency bands allocated to BSS for different regions of the world.

As mentioned earlier, one of the major factors affecting the quality of service in BSS/DBS is the attenuation of the received signal due to rain especially at higher frequencies where the impact is more pronounced than at lower frequencies. ITU has carried out extensive studies in this area, based on the ITU Radiocommunication Assembly's recommendation which emphasized the need for collecting statistical data of precipitation intensity for the prediction of attenuation and scattering caused by precipitation. Based on this recommendation, Table 3.23 can be used to obtain the expected median cumulative distribution of rain rate for the rain climate region, and Figure 3.84 can be used to select the rain climate region for the prediction of precipitation effects.

In the following, the original copy of the paper published in 1945 by A. C. Clarke in Wireless World Journal, as was referred to earlier in this chapter, is presented due to its importance and relevance to the topic of our discussion.

TABLE 3.23

Rainfall Intensity Exceeded (mm/h)

Percentage of Time (%)	A	B	C	D	E	F	G	H	J	K	L	M	N	P	Q
1.0	<0.1	0.5	0.7	2.1	0.6	1.7	3	2	8	1.5	2	4	5	12	24
0.3	0.8	2	2.8	4.5	2.4	4.5	7	4	13	4.2	7	11	15	34	49
0.1	2	3	5	8	6	8	12	10	20	12	15	22	35	65	72
0.03	5	6	9	13	12	15	20	18	28	23	33	40	65	105	96
0.01	8	12	15	19	22	28	30	32	35	42	60	63	95	145	115
0.003	14	21	26	29	41	54	45	55	45	70	105	95	140	200	142
0.001	22	32	42	42	70	78	65	83	55	100	150	120	180	250	170

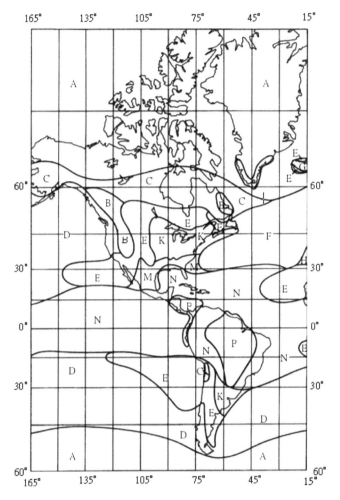

FIGURE 3.84

PN.837-11, Recommendation ITU-R Pn.837-1 characteristics of precipitation for propagation modelling. https://slidept.net/document/rec-itu-r-pn-837-1-1-recommendation-itu-r-pn-837-1-characteristics-of-precipitation-for-propagation-modelling.[77]

Extra-Terrestrial Relays: Can Rocket Stations Give World-Wide Radio Coverage?

By Arthur C. Clarke (*Wireless World*, October 1945, 305–308.)

Although it is possible, by a suitable choice of frequencies and routes, to provide telephony circuits between any two points or regions of the earth for a large part of the time, long distance communication is greatly hampered by the peculiarities of the ionosphere, and there are even occasions when it may be impossible. A true broadcast service, giving constant field strength at all times over the whole globe would be invaluable, not to say indispensable, in a world society.

Unsatisfactory though the telephony and telegraph position is, that of television is far worse, since ionospheric transmission cannot be employed at all. The service area of a television station, even on a very good site, is only about a 100 miles across. To cover a small country such as Great Britain would require a network of transmitters, connected by coaxial lines, waveguides or Very High Frequency (VHF) relay links. A recent theoretical study[1] has shown that such a system would require repeaters at intervals of 50 miles or less. A system of this kind could provide television coverage, at a very considerable cost, over the whole of a small country. It would be out of the question to provide a large continent with such a service, and only the main centers of population could be included in the network.

The problem is equally serious when an attempt is made to link television services in different parts of the globe. A relay chain several thousand miles long would cost millions, and transoceanic services would still be impossible. Similar considerations apply to the provision of wide-band frequency modulation and other services, such as high-speed facsimile which are by their nature restricted to the ultra-high-frequencies.

Many may consider the solution proposed in this discussion too far-fetched to be taken very seriously. Such an attitude is unreasonable, as everything envisaged here is a logical extension of developments in the last 10 years—in particular, the perfection of the long-range rocket of which V2 was the prototype. While this article was being written, it was announced that the Germans were considering a similar project, which they believed possible within 50–100 years.

Before proceeding further, it is necessary to discuss briefly certain fundamental laws of rocket propulsion and "astronautics." A rocket which achieved a sufficiently great speed in flight outside the earth's atmosphere would never return. This "orbital" velocity is 8 km/s (5 miles per second), and a rocket which attained it would become an artificial satellite, circling the world forever with no expenditure of power—a second moon, in fact. The German transatlantic rocket A10 would have reached more than half this velocity.

It will be possible in a few more years to build radio-controlled rockets which can be steered into such orbits beyond the limits of the atmosphere and left to broadcast scientific information back to the Earth. A little later, manned rockets will be able to make similar flights with sufficient excess power to break the orbit and return to Earth.

There are an infinite number of possible stable orbits, circular and elliptical, in which a rocket would remain if the initial conditions were correct. The velocity of 8 km/s applies only to the closest possible orbit, one just outside the atmosphere, and the period of revolution would be about 90 minutes. As the radius of the orbit increases the velocity decreases, since gravity is diminishing and less centrifugal force is needed to balance it. Figure 1 shows this graphically. The moon, of course, is a particular case and would lie on the curves of Figure 1 if they were produced. The proposed German space-stations would have a period of about four and a half hours.

It will be observed that one orbit, with a radius of 42,000 km, has a period of exactly 24 hours. A body in such an orbit, if its plane coincided with that of the Earth's equator, would revolve with the Earth and would thus be stationary above the same spot on the planet. It would remain fixed in the sky of a whole hemisphere and unlike all other heavenly bodies would neither rise nor set. A body in a smaller orbit would revolve more quickly than the Earth and so would rise in the west, as indeed happens with the inner moon of Mars.

Using material ferried up by rockets, it would be possible to construct a "space-station" in such an orbit. The station could be provided with living quarters, laboratories, and everything needed for the comfort of its crew, who would be relieved and provisioned by a regular rocket service. This project might be undertaken for purely scientific reasons as it would contribute enormously to our knowledge of astronomy,

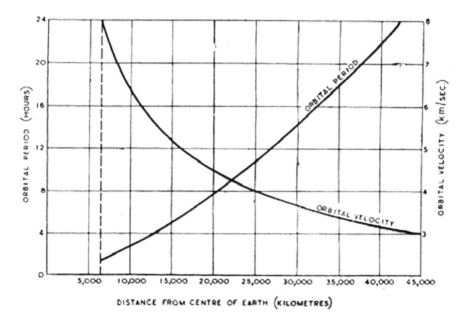

FIGURE 1
Variation of orbital period and velocity with distance from the center of the earth.

physics, and meteorology. A good deal of literature has already been written on the subject.[2]

Although such an undertaking may seem fantastic, it requires for its fulfillment rockets only twice as fast as those already in the design stage. Since the gravitational stresses involved in the structure are negligible, only the very lightest materials would be necessary and the station could be as large as required.

Let us now suppose that such a station were built in this orbit. It could be provided with receiving and transmitting equipment (the problem of power will be discussed later) and could act as a repeater to relay transmissions between any two points on the hemisphere beneath, using any frequency which will penetrate the ionosphere. If directive arrays were used, the power requirements would be very small, as direct line of sight transmission would be used. There is the further important point that arrays on the Earth, once set up, could remain fixed indefinitely.

Moreover, a transmission received from any point on the hemisphere could be broadcast to the whole of the visible face of the globe, and thus the requirements of all possible services would be met (Figure 2).

It may be argued that we have as yet no direct evidence of radio waves passing between the surface of the Earth and outer space; all we can say with certainty is that the shorter wavelengths are not reflected back to the Earth. Direct evidence of field strength above the Earth's

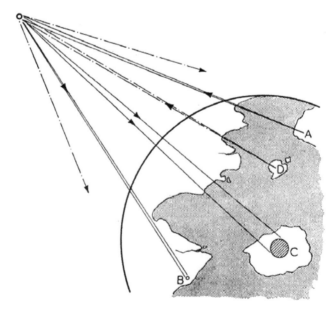

FIGURE 2
Typical extra-terrestrial relay services. Transmission from A being relayed to point B and area C; transmission from D being relayed to whole hemisphere.

atmosphere could be obtained by V2 rocket technique, and it is to be hoped that someone will do something about this soon as there must be quite a surplus stock somewhere! Alternatively, given sufficient transmitting power, we might obtain the necessary evidence by exploring for echoes from the moon. In the meantime, we have visual evidence that frequencies at the optical end of the spectrum pass through with little absorption except at certain frequencies at which resonance effects occur. Medium high frequencies go through the E layer twice to be reflected from the F layer and echoes have been received from meteors in or above the F layer. It seems fairly certain that frequencies from, say, 50–100,000 Mc/s could be used without undue absorption in the atmosphere or the ionosphere.

A single station could only provide coverage to half the globe, and for a world service three would be required, though more could be readily utilized. Figure 3 shows the simplest arrangement. The stations would be arranged approximately equidistantly around the Earth, and the following longitudes appear to be suitable:

The stations in the chain would be linked by radio or optical beams, and thus any conceivable beam or broadcast service could be provided.

The technical problems involved in the design of such stations are extremely interesting,[3] but only a few can be gone into here. Batteries of parabolic reflectors would be provided, of apertures depending on the frequencies employed. Assuming the use of 3,000 Mc/s waves, mirrors about a meter across would beam almost all the power on to the Earth. Larger reflectors could be used to illuminate single countries or regions for the more restricted services, with consequent economy of power. On the higher frequencies it is not difficult to produce beams less than a degree in width, and, as mentioned before, there would be no physical limitations on the size of the mirrors. (From the space station, the disc of the Earth would be a little over 17° across.) The same mirrors could be used for many different transmissions if precautions were taken to avoid cross-modulation.

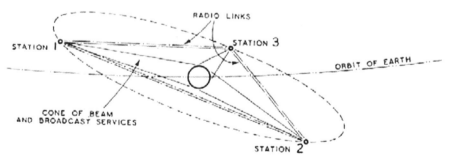

FIGURE 3
Three satellite stations would ensure complete coverage of the globe.

It is clear from the nature of the system that the power needed will be much less than that required for any other arrangement, since all the energy radiated can be uniformly distributed over the service area, and none is wasted. An approximate estimate of the power required for the broadcast service from a single station can be made as follows: The field strength in the equatorial plane of a $\lambda/2$ dipole in free space at a distance of d metres is[4]

$$e = 6.85\frac{Vp}{d} \text{ V/m,}$$

where P is the power radiated in watts.

Taking d as 42,000 km (effectively it would be less), we have $P = 37.6\ e^2$ W (e now in μV/m.)

If we assume e to be 50 μV/m, which is the F.C.C. standard for frequency modulation, P will be 94 kW. This is the power required for a single dipole, and not an array which would concentrate all the power on the Earth. Such an array would have a gain over a simple dipole of about 80. The power required for the broadcast service would thus be about 1.2 kW.

Ridiculously small though it is, this figure is probably much too generous. Small parabolas about a foot in diameter would be used for receiving at the Earth end and would give a very good signal noise ratio. There would be very little interference, partly because of the frequency used and partly because the mirrors would be pointing toward the sky which could contain no other source of signal. A field strength of 10 μV/m might well be ample, and this would require a transmitter output of only 50 W.

When it is remembered that these figures relate to the broadcast service, the efficiency of the system will be realized. The point-to-point beam transmissions might need powers of only 10 W or so. These figures, of course, would need correction for ionospheric and atmospheric absorption, but that would be quite small over most of the band. The slight falling off in field strength due to this cause toward the edge of the service area could be readily corrected by a non-uniform radiator.

The efficiency of the system is strikingly revealed when we consider that the London Television service required about 3 kW average power for an area less than 50 miles in radius.[5]

A second fundamental problem is the provision of electrical energy to run the large number of transmitters required for the different services. In space beyond the atmosphere, a square meter normal to the solar radiation intercepts 1.35 kW of energy.[6] Solar engines have already been devised for terrestrial use and are an economic proposition in tropical countries. They employ mirrors to concentrate sunlight on the boiler of a low-pressure steam engine. Although this arrangement is not very

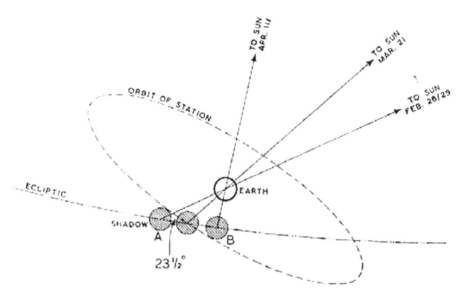

FIGURE 4
Solar radiation would be cut off for a short period each day at the equinoxes.

efficient it could be made much more so in space where the operating components are in a vacuum, the radiation is intense and continuous, and the low-temperature end of the cycle could be not far from absolute zero. Thermo-electric and photoelectric developments may make it possible to utilize the solar energy more directly.

Though there is no limit to the size of the mirrors that could be built, 150 m in radius would intercept over 10,000 kW and at least a quarter of this energy should be available for use.

The station would be in continuous sunlight except for some weeks around the equinoxes, when it would enter the Earth's shadow for a few minutes every day. Figure 4 shows the state of affairs during the eclipse period. For this calculation, it is legitimate to consider the Earth as fixed and the Sun as moving round it. The station would graze the Earth's shadow at A, on the last day in February. Every day, as it made its diurnal revolution, it would cut more deeply into the shadow, undergoing its period of maximum eclipse on March 21 on that day it would only be in darkness for 1 hour 9 minutes. From then onwards the period of eclipse would shorten, and after April 11 (B) the station would be in continuous sunlight again until the same thing happened 6 months later at the autumn equinox, between September 12 and October 14. The total period of darkness would be about 2 days per year, and as the longest period of eclipse would be little more than an hour there should be no difficulty in storing enough power for an uninterrupted service.

Conclusion

Briefly summarized, the advantages of the space station are as follows:

1. It is the only way in which true world coverage can be achieved for all possible types of service.
2. It permits unrestricted use of a band at least 100,000 Mc/s wide, and with the use of beams an almost unlimited number of channels would be available.
3. The power requirements are extremely small since the efficiency of "illumination" will be almost 100%. Moreover, the cost of the power would be very low.
4. However great the initial expense, it would only be a fraction of that required for the world networks replaced, and the running costs would be incomparably less.

Appendix—Rocket Design

The development of rockets sufficiently powerful to reach "orbital" and even "escape" velocity is now only a matter of years. The following figures may be of interest in this connection.

The rocket has to acquire a final velocity of 8 km/s. Allowing 2 km/s for navigational corrections and air resistance loss (this is legitimate as all space rockets will be launched from very high country) gives a total velocity needed of 10 km/s. The fundamental equation of rocket motion is[2]

$$V = v\log_e R$$

where V is the final velocity of the rocket, v the exhaust velocity and R the ratio of initial mass to final mass (payload plus structure). So far v has been about 2–2.5 km/s for liquid fuel rockets but new designs and fuels will permit of considerably higher figures. (Oxyhydrogen fuel has a theoretical exhaust velocity of 5.2 km/s and more powerful combinations are known.) If we assume v to be 3.3 km/s. R will be 20 to I. However, owing to its finite acceleration, the rocket loses velocity as a result of gravitational retardation. If its acceleration (assumed constant) is a m/s,[2] then the necessary ratio Rg is increased to

$$R_g = R\frac{a+g}{a}$$

For an automatically controlled rocket a would be about 5 g and so the necessary R would be 37 to I. Such ratios cannot be realised with a single rocket but can be attained by "step-rockets,"[2] while very much higher ratios (up to 1,000 to i) can be achieved by the principle of "cellular construction."[3]

Epilogue—Atomic Power

The advent of atomic power has at one bound brought space travel half a century nearer. It seems unlikely that we will have to wait as much as 20 years before atomic-powered rockets are developed, and such rockets could reach even the remoter planets with a fantastically small fuel/mass ratio—only a few percent. The equations developed in the appendix still hold, but v will be increased by a factor of about a thousand.

In view of these facts, it appears hardly worthwhile to expend much effort on the building of long distance relay chains. Even the local networks which will soon be under construction may have a working life of only 20–30 years.

References

1. Radio-relay systems, C. W. Hansell. Proc. I.R.E., Vol. 33, March, 1945.
2. Rockets, Willy Ley. (Viking Press, New York).
3. Das Problem der Befahrung des Weltraums, Hermann Noordung.
4. Frequency modulation, A. Hund. (McGraw Hill).
5. London Television Service, MacNamara and Birkinshaw. J.I.E.E., December 1938.
6. The Sun, C.G. Abbot. (Appleton-Century Co.).
7. Journal of the British Interplanetary Society. January 1939.

3.3 Questions

1. Name the different orbits used in satellite communications and briefly explain the characteristics of each orbit.

2. Name the three types of LEO satellite systems used in satellite communications and explain the followings for each orbit:
   - The frequency bands that they operate at.
   - Services that they provide.
   - Example of a satellite system designed or operating in each case.

3. The frequency bands most commonly used in satellite communication are C-band, K_u-band, and K_a-band. Compare these frequency bands in terms of attenuation, bandwidth, interference with terrestrial systems, and Earth station antenna size.

4. Draw the diagram for different modes of interferences between a terrestrial station and two satellite systems (composed of the Earth station and space segment) and explain the possible inference modes.

5. Define TT&C and explain the function of each element on the ground and on the spacecraft.

6. What are the different kinds of power amplifiers used in satellite communication? Name them, explain each one in terms of power capabilities, and back-off factor.

7. Show and explain the effect of wide-beam and narrow-beam Earth station antennas on GEO arc utilization.

8. Name the three main categories of satellite communication services according to ITU and briefly explain each one.

9. Draw the Center-Feed antenna and Offset-Feed antenna and explain which one is preferable and what are the limiting factors for using the Offset-Feed antenna.

10. Define VSAT, its topologies, and briefly explain some of its applications.

11. Name the subsystems onboard of a communication satellite and briefly describe the function of each of the subsystems.

12. Explain different methods of making more efficient use of the GEO.

13. As of December 2015, how many satellites were operational and what were their functions?

14. Name different types of commercial satellites.

15. Name some of the satellite's benefits.

16. Explain the three Kepler's laws.

17. What are the three remote-sensing platforms used to collect information from ground-based targets and compare them in terms of coverage area and resolution.

18. Describe the active and passive remote sensing and give examples for each one.

19. Within the electromagnetic spectrum, what are the two bands that most of the scientific research in remote sensing carries out? Explain each band's characteristics.

20. Explain SAR and its significance in satellite remote sensing.

21. Name the ground-surface characteristics affecting the EM signal reflection.

22. Explain the high, medium, and low resolutions remote sensing and name satellites that provide each of the above.

23. What is the major difference between thermal infrared remote sensing (3–$100\,\mu m$) and visible and near infrared (0.4–$2.5\,\mu m$) based remote sensing.

24. Explain the EIRP relationship with the antenna size on the ground and define spot, zonal, regional, hemispheric, and global beam coverage and their corresponding EIRP.

25. Explain what Steerable phased-array antenna is and what its main applications are.
26. What are the functions of Earth station?
27. Why uplink frequency is higher than downlink frequency?
28. What are the applications of VSAT network?
29. Define Transponder and explain its function.
30. Write link equation expression in decibels.
31. What is the difference between satellite Earth station and microwave link?
32. Explain about launch and launch-vehicles for the communication satellite.
33. Explain how to find position location in GPS system.
34. For which satellite communications link, the frequency of 40–60 GHz is used?
35. Within the RF spectrum, identify the shared frequency bands with terrestrial systems and their designations.
36. Explain the types of satellite stabilization.
37. Name the three main building blocks of any satellite communication system.
38. What are the main shortfalls of geostationary satellite communication systems and explain if there are any possible solutions to overcome these shortfalls.

3.4 Problems

1. Calculate the average orbital velocity of the Earth given that:

   $r = 149.6 \times 106$ km (radius of the orbit of the Earth)

   $p = 365.25$ days (time required for the Earth to complete one orbit around the Sun)

   $C = 2\pi r$ (circumference of a circle)

   $\pi = 3.14$

   Express the velocity in km/s

2. A satellite is in an 800 km high circular orbit. Determine:

   a. The orbital angular velocity in radians per second
   b. The orbital period in minutes; and
   c. The orbital velocity in meters per second.

3. A C-band Earth station has an antenna with a transmit gain of 56 dB. The transmitter output power is set to 110 W at a frequency of 6.00 GHz. The signal is received by a satellite at a distance of 36,500 km by an antenna with a gain of 25 dB. The signal is then routed to a transponder with a noise temperature of 480 K, a bandwidth of 36 MHz, and a gain of 105 dB.

   a. Calculate the path loss at 6.0 GHz. Wavelength is 0.05 m.

   b. Calculate the power at the output port of the satellite antenna, in dBW.

   c. Calculate the noise power at the transponder input, in dBW, in a bandwidth of 36 MHz.

   d. Calculate the C/N ratio, in dB, in the transponder.

   e. Calculate the carrier power, in dBW and in watts, at the transponder output.

4. Geostationary satellites use C, K_u, and K_a bands. The path length from an Earth station to the GEO satellite is 36,000 km. For this range, calculate the path loss in decibels for the following frequencies:

   a. 6.1 GHz, 4.2 GHz

   b. 14.2, GHz, 12.1 GHz

   c. 30 GHz, 20 GHz

5. A geostationary satellite carries a transponder with a 25 W transmitter at 4 GHz. The transmitter is operated at an output power of 12 W and drives an antenna with a gain of 32 dB.

   An Earth station is at the center of the coverage zone of the satellite, at a range of 36,000 km find:

   a. The flux density at the Earth station in dBW/m²

$$\left(F = 20\log\left[P_t G_t/\left(4\pi R^2\right)\right] \text{dBW/m}^2 \right)$$

   b. The power received by an antenna with a gain of 40 dB, in dBW.

   c. The EIRP of the transponder in dBW.

6. A satellite in GEO orbit is a distance of 36,000 km from an Earth station. The required flux density at the satellite to saturate one transponder at a frequency of 14.1 GHz is −88.0 dBW/m². The Earth station has a transmitting antenna with a gain of 50 dB at 14.1 GHz. Find:

   a. The EIRP of the earth station

   b. The output power of the Earth station transmitter.

7. A 4 GHz Earth station receiving system has an antenna with a noise temperature of 40 K, a LNA with a noise temperature of 80 K and a gain of 38 dB, and a mixer with a noise temperature of 900 K. Find the system noise temperature.

8. A Direct Broadcast Television satellite is in GEO at 100° west longitude. It carries 16 transponders, each with a saturated output power of 180 W and a bandwidth of 26 MHz. The antenna on the satellite has a gain of 32 dB. The receiving terminals all use antennas with a circular aperture with a diameter of 20 in. and an aperture efficiency of 60%. The noise bandwidth of the digital TV receiver is 20 MHz. Use a distance to the GEO satellite of 36,500 km in your calculations.

   a. Calculate the free space path loss and the receiving terminal antenna gain at 12 GHz.

   b. The receiving terminal has a system noise temperature of 100 K in clear air. Draw up a noise power budget for the receiver using the receiver's noise bandwidth.

   c. Calculate the clear air C/N ratio for the receiver with a noise bandwidth of 20 MHz.

      The minimum permissible C/N ratio is 10.0 dB. What is the clear air link margin?

References

1. NASA. What Is a Satellite? www.nasa.gov/audience/forstudents/k-4/stories/what-is-a-satellite-k4.html.
2. United Nations Office for Outer Space Affairs (UNOOSA).
3. The Union of Concerned Scientists (UCS).
4. Switzerland Janitor Satellites, www.theguardian.com/science/2012/feb/15/swiss-create-janitor-satellite-space cleanup.
5. The TAURI Group 2016. State of the Satellite Industry Report. Satellite Industry Association, Retrieved on November 2016, www.sia.org/wp-content/uploads/2016/06/SSIR16-Pdf-Copy-for-Website-Compressed.pdf.
6. NESDIS. Satellites. Retrieved on July 4, 2008, https://en.wikipedia.org/wiki/Weather_satellite.
7. NOAA. NOAA Satellites, Scientists Monitor Mt. St. Helens for Possible Eruption. Retrieved on July 4, 2008.
8. "U.S. Launches Camera Weather Satellite". *The Fresno Bee. AP and UPI*. April 1, 1960. pp. 1a, 4a.
9. National Environmental Satellite Center. January 1970. SIRS and the improved marine weather forecast. *Marners Weather Log*, 14(1): 12–15. Environmental Science Services Administration.
10. Labrador, V. 2015. "Satellite Communication". Britannica.com. Retrieved 2016 02–10.
11. "Satellites – Communication Satellites". Satellites.spacesim.org. Retrieved 2016-02-10.
12. "Military Satellite Communications Fundamentals | The Aerospace Corporation". *Aerospace*. 2010-04-01. Retrieved 2016-02–10.

13. ITU Radio Regulations, Section IV. Radio Stations and Systems – Article 1.39, definition: Broadcasting-satellite service.
14. "Frequency Letter Bands". *Microwaves101.com*. 25 April 2008.
15. "Installing Consumer-Owned Antennas and Satellite Dishes". FCC. Retrieved 2008-11-21.
16. www.britannica.com/topic/spaceflight#ref741344.
17. "Galileo Goes Live!" *europa.eu*. 2016-12-14.
18. "Beidou Satellite Navigation System to Cover Whole World in 2020". Eng.chinamil.com.cn. Retrieved 2011-12-30.
19. "DubaiSat-2, Earth Observation Satellite of UAE". *Mohammed Bin Rashid Space Centre*.
20. "DubaiSat-1, Earth Observation Satellite of UAE". *Mohammed Bin Rashid Space Centre*.
21. Root, J. 2004. Orbital Speed, www.freemars.org/jeff/speed/index.htm.
22. www.satcom.co.uk/print.asp?article=29.
23. Roddy, D. 2001. *Satellite Communication*, 3rd Edition. New York: McGraw-Hill Telecom Engineering.
24. Space Certification Program, Satellite Tool Kit, www.spaceconnection.org/certified-products/satellite-tool-kit.
25. AGI Analysis Software for Land, Sea, Air, and Space, January 2008.
26. www.spacesafetymagazine.com/space-on-earth/space-economy/.
27. THE ECONOMICS OF SATELLITE COMMUNICATIONS. Marcellus Snow, Associate Professor of Economics, University of Hawaii at Manoa.
28. Lillesand, T.M., and R.W. Kiefer. 1999. *Remote Sensing and Image Interpretation*. Chichester: Wiley.
29. Fritz, L.W. High Resolution Commercial Remote Sensing Satellites and Spatial Information Systems, p. 5, www.isprs.org/.
30. http://unfccc.int/essential_background/items/6031.php.
31. http://copernicus.eu/main/towards-european-operational-observing-system-monitor-fossil-co2-emissions.
32. www.oneonta.edu/faculty/baumanpr/geosat2/RS%20History%20II/RS-History-Part-2.html. https://en.wikipedia.org/wiki/Nimbus_progra.
33. www.e-education.psu.edu/natureofgeoinfo/c8_p12.html.
34. www.seos-project.eu/modules/remotesensing/remotesensing-c03-p02.html.
35. "NASA: TERRA (EOS AM-1)". *nasa.gov*. Retrieved 2011-01–07.
36. Maurer, J. 2001. "Overview of NASA's Terra Satellite". hawaii.edu (University of Hawai'i). Retrieved 2011-01-07.
37. Stevens, N.F., H. Garbeil, P.J. Mouginis-Mark. 2004. "NASA EOS Terra ASTER: Volcanic Pographic Mapping and Capability" (PDF). Hawai'i Institute of Geophysics and Planetology. Retrieved 2011-01-07.
38. "MODIS Terra Satellite Images". ucar.edu (National Center for Atmospheric Research: Earth Observatory Laboratory). Retrieved 2011-01-07.
39. "MEASUREMENTS OF POLLUTION IN THE TROPOSPHERE (MOPITT)". acd.ucar.edu (NESL's Atmospheric Chemistry Division). Archived from the original on 2011-01-28. Retrieved 2011-01–07.
40. "NASA's Terra Satellite Tracks Global Pollution". gsfc.nasa.gov (Goddard Space Flight Center). 2004-05-18. Archived from the original on 2010-11-24. Retrieved 2011-01–07.
41. https://eospso.nasa.gov/.

42. www.daviddarling.info/encyclopedia/A/ADEOS.html.
43. https://directory.eoportal.org/web/eoportal/satellite-missions/r/radarsat-2.
44. www.n2yo.com/satellite/?s=40894.
45. www.isro.gov.in/Spacecraft/cartosat-2-series-satellite-2.
46. www.pinterest.com/pin/456622849700758499/.
47. https://en.wikipedia.org/wiki/Geostationary_Operational_Environmental_Satellite.
48. www.hughesscgheritage.com/2015/08/.
49. http://spaceflight101.com/falcon-9-jcsat-14/jcsat-14-satellite/.
50. http://ww2010.atmos.uiuc.edu/(Gh)/guides/rs/sat/goes/curg.rxml.
51. www.eumetsat.int/website/home/Satellites/CurrentSatellites/Meteosat/MeteosatDesign/index.html.
52. Kishtawal, C.M. Meteorological Satellites, Atmospheric Science Division, Meteorology and Oceanographic Group, Space Application Center (ISRO), Ahmedabad, India.
53. Dana, P.H. Global Positioning System Overview, www.colorado.edu/geography/gcraft/notes/gps/gps_f.html.
54. https://commons.wikimedia.org/wiki/File:Mercury_Spacecraft.png.
55. Apollo Diagrams, http://history.nasa.gov/SP-4225/diagrams/apollo/apollo-diagram.htm.
56. Soviet Web Space, Sputnik: A History of the Beginning of the Space Age, http://faculty.fordham.edu/siddiqi/sws/sputnik/sputnik.html.
57. Clarke, A.C. 1945. Extra-terrestrial relays, can rocket stations give world-wide radio coverage? *Wireless World*, 51: 305–308.
58. En. Fahrul Hakim Ayob et al. 2003. Very Small Aperture Terminal, July 22.
59. http://allaboutshimoga.blogspot.com/2009/11/satellite-transponder-communications.html.
60. Encyclopædia Britannica, Inc., www.britannica.com/topic/spaceflight#ref741344.
61. Keesee, J. Satellite Communications, ocw.mit.edu/courses/aeronauticsand-astronautics/16-851-satellite-engineering-fall-2003/lecturenotes/l21satelite-comm2_done.pdf.
62. Fixed-Satellite Service Plan (AP30B), Radiocommunication Bureau, IT, Spectrum Management System for Developing Countries, Frequency Allocation Table, www.itu.int/ITU-D/asp/CMS/Events/2010/SMS4DC/SMS4DC2_AllocationsV2.pdf.
63. www.ictp.it/~pub_off/lectures/lns016/Vol_16_Annex.pdf.
64. CRASA Wireless Technologies Policy and Regulation, September 2004, www.crasa.org/download.php?doc=doc_pub_eng62.pdf.
65. Pritchard, W., L.H. Suyderhoud, and R.A. Nelson. 1993. *Satellite Communication Systems Engineering*, 2nd Edition, Upper Saddle River, NJ: Prentice Hall PTR.
66. Applied Aerospace Structures Corp, AASC. (Specializing in design, fabrication, and testing of lightweight aerospace structural assemblies.) Stockton, California.
67. www.esa.int/Our_Activities/Telecommunications_Integrated_Applications/Satellite_frequency_bands.
68. http://eie.uonbi.ac.ke/sites/default/files/cae/engineering/eie/RADIO%20FREQUENCY%20INTERFERENCE%20IN%20SATELLITE%20COMMUNICATIONS%20SYSTEM.pdf.
69. https://www.commsmea.com/17445-evolution-of-satellite-communications.

70. The Comsys VSAT Report, 11th Edition, VSAT Network Types, www.comsys. co.uk/wvr_nets.htm.
71. www.intelsatgeneral.com/satellite-coverage-maps/.
72. www.itso.int/dyn4000/itso/tpl1_itso.cfm?location=&id=338&link_ src=HPL&lang=english.
73. https://en.wikipedia.org/wiki/List_of_Intelsat_satellites.
74. Intelsat Earth Station Technology, Revision 5, June 1999.
75. www.navtec.de/english/inmarsat.htm.
76. http://encyclopedia2.thefreedictionary.com/Direct+broadcasting+satellite+ systems.
77. Rec. ITU-R PN.837-1 1, Recommendation ITU-R Pn. 837-1 Characteristics of precipitation for propagation modelling.

4

Future Space Technologies

4.1 Introduction

Future space technologies could be viewed from those organizations that are in the forefront of space technologies. With no doubt, the National Aeronautics and Space Administration (NASA) would be on the top of such a list and their dreams and imagination for the years and perhaps decades ahead would give us a rather good understanding of where we are moving toward and what the goals and targets are in this area. Space technology has affected the lives of millions in different ways as we learned in this book so far, from numerous applications. The impact on education, safety and security, early warnings of natural disasters, agriculture, city planning, navigation, communication, and tens, perhaps hundreds of other areas that our lives have been affected, are only some examples of the applications of such technologies. We experience some of these applications on a day-to-day basis, and they have become part of our lives in such a way that we do not realize them. Some of the immediate impacts are hidden in research activities in the space environment that gradually find their ways into different disciplines. Although one might think that curiosity was perhaps the primary incentive for humans to search space and beyond, and of course no one denies that, at the same time we should realize that space-related activities are now taking a different shape and direction. Maybe economy is part of the reason that we no longer can afford to go to space just for the sake of going to space and partly because the conquest of the space in the last millennium has been sufficient to satisfy our ego and desire of wanting to go to space. Now we are at a different time when we go to space for a reason that is not only economically justifiable, but we can see its immediate impact on our lives whether the impact is on education, the environment, peace and security, agriculture, or the advancement of science and technology which paves the way for a better future for generations to come. To understand the scope of such impact, we will look at some of the emerging space companies that have created a new landscape of the space exploration.

Today's space companies aim to develop viable businesses to provide services such as human space travel and habitation—currently the domain of government agencies. The re-emergence of private investment in space, as we have witnessed in the last few years, coupled with ongoing government investment, combines for a promising future in space exploration. A belief in space exploration and economic development drives the current wave of space entrepreneurs, who seek revolutionary change in the humans' relationship with space, and are motivated by their missions. They do not necessarily view exploration as movement toward one singular accomplishment (such as a moon landing), but as the development of sustainable and diverse economic activities in space. This requires new capabilities enabled by new technologies, but also existing capabilities (like access to space) available at lower costs and prices. By developing these products and services, they aim to open new markets for space experiences and industrial activity. The following is an overview of emerging space companies and their visions of space exploration. Many of these companies are entrepreneurial, but several (Sierra Nevada, Boeing, and ULA) are accomplished aerospace firms seeking to drive new markets. In addition, there are many other emerging space companies in supplier, component, and services segments. This is a dynamic industry with frequent changes and the listing here is only a sample of the diverse efforts under way. This is not an exhaustive list and several of these companies plan to provide more than one type of service, for example, in addition to providing orbital launch services, SpaceX also intends to provide crew and cargo launch services to Mars sometime in the 2030s.

	Company	Vehicle(s) or Spacecraft	Services
Space Access	Blue Origin	New Shepard, Biconic Spacecraft	Suborbital and orbital launch services including human spaceflight
	Masten Space Systems	Xaero, Xogdor	Suborbital launches of small payloads
	Virgin Galactic	SpaceShipTwo, LauncherOne	Suborbital launches of small payloads; suborbital human spaceflight, and air-launched nanosatellite launches
	XCOR Aerospace	Lynx	Suborbital launches of small payloads, suborbital human spaceflight, and nanosatellite launches
	Orbital Sciences Corporation	Pegasus, Tauris, Antares, Cygnus	Orbital launches of satellites and ISS cargo
	SpaceX	Falcon 9, Falcon Heavy, Dragon	Orbital launches of satellites and ISS cargo, with orbital human spaceflight planned by 2017
	Stratolaunch Systems	Stratolauncher	Air-launched orbital launch services
	United Launch Alliance	Atlas V, Delta IV	Orbital launch services
Remote Sensing	Planet Labs	Dove, Flock 1	Frequent imaging of the Earth and open access to acquired data via website
	Skybox Imaging	SkySat	Frequent imaging and HD video of the Earth, data analysis, and open access to acquired data via website
LEO Human Spaceflight	Bigelow Aerospace	BA 330	Inflatable habitats for use in orbit or on the Moon
	Boeing	CST-100	Crewed LEO transportation
	Sierra Nevada Corporation	Dream Chaser	Crewed LEO transportation
	Space Adventures	Soyuz	Crewed LEO and lunar expeditions
Beyond LEO	B612 Foundation	Sentinel	Detection and characterization of potentially hazardous asteroids
	Inspiration Mars Foundation	Inspiration Mars	Crewed Mars flyby expedition
	Moon Express	Moon Express	Prospecting and mining lunar resources
	Planetary Resources	Arkyd 100, Arkyd 200, Arkyd 300	Prospecting and mining asteroid resources

List of emerging space companies, grouped by destination.

Source: **The Evolving Landscape of 21st Century American Spaceflight, www.nasa.gov.**

As was mentioned above, one of the reasons for the expansion of the space technology is the economy aspects of it, which will be briefly, touched up in the followings:

Emerging Space in 2044

Where are all these diverse and energetic emerging space activities leading us to in the future? The President has directed that by the mid-2030s NASA's human exploration wave front will have reached Mars with a human mission to an asteroid as a stepping-stone. Private sector entities and other nations are presently planning new stations, missions to the Moon, and even Mars. Robotic missions will continue throughout the solar system, featuring a mix of large, complex spacecraft and small but sophisticated micro-satellites and probes. Humanity stand today at the beginning of a grand sequence of voyages into the cosmos. NASA is on a path of exploration and pioneering in the inner solar system, opening new opportunities and challenges for future generations and driving the American economic expansion into space. Although the specific shape and rate of this economic expansion depends as much on the actions of the individuals, corporations, and foundations involved as it does on NASA, the following descriptions present one view of the future in 2044 as it could be as NASA catalyzes an expansion of American space activities through its programs and partnerships.

The following is NASA's vision on the next steps out into the solar system. This outline builds on the framework and philosophy described in the NASA's report

Voyages: Charting the Course for Sustainable Human Space Exploration.[1]

Although the order of presentation represents present prioritization, all three sets of destinations are expected to be critical parts of space development in the decades to come and we should expect them to take different levels of prioritization at different times depending on political, economic, and technological circumstances.

- **Near-Earth Asteroids**
  NASA's near-earth asteroid explorations paved the way for regular government and private sector activity involving near-earth asteroids. With the support of NASA partnerships, companies began extracting water ice, metals, and rock from near-Earth asteroids (NEAs) identified as viable candidates for mining, beginning the process of providing materials and propellant for the expanding interplanetary transportation system and economy. As on the Moon, activities relating to mining necessitated the development of more robust reusable systems capable of routinely operating in harsh environments. NASA's human exploration of NEAs contributed

insights valuable both to governments and private sector actors regarding NEA characteristics, proximity operations, extraction techniques, and options for planetary defense from the threat of asteroid impact.

- **Mars**
  As the greatest challenge of American space exploration of the age, the American private sector mustered significant expenditure and investment in order to advance the date of this achievement in partnership with NASA and enabled long-term habitation following NASA's initial missions. Martian surface asset emplacement activities occurred periodically to test systems and deliver cargo for the upcoming landing, include supplemental private sector activities. Prior to the first human steps on Mars, NASA and privately financed expeditions traveled to Mars orbit and visited Phobos to validate systems in preparation for the historic expedition to the Martian surface.

- **Earth and Moon**
  The International Space Station continues to serve as a venerable space research and development institution into the 2020s when the first commercial modular stations began to be deployed for microgravity applications and private sector expeditions. Reductions in launch costs, the emergence of low-cost satellite development kits, and the introduction of commercial spaceflight opportunities allowed more people than ever before to have direct access to space—inspiring and training a resurgent generation of aerospace engineers, scientists, and technologists in the process. Space traffic management and active debris removal and mitigation, topics that are discussed in different sections of the present book, were developed to address the greater amount of orbital activity. Propellant depots and spacecraft servicing systems located in GEO and Lagrangian Libration Points provided support for the growing interplanetary transportation system. Some of these facilities were established by governments as independent authorities that could operate independently and could raise their own funds and issue their own bonds, as do seaports on Earth. These served as gateways to points beyond as cargo and crew spacecraft launched from Earth prepared for journeys to the Moon, asteroids, and Mars. The Moon and cis-lunar environment became a primary proving-ground for exploration systems and technology development, particularly for the development of robotic planetary surface systems. Surface operations were supported by a modest station, communications satellites, navigation satellites, and remote sensing satellites in lunar orbit. Prospecting,

extraction, and processing of lunar volatiles and metals began to scale up to support activities on the Moon as well as to sell propellant for in-space transportation. The costs of working on the Moon fell due to the investments in space transportation infrastructure and local production. Exploration of the lunar surface was supplemented by privately financed expeditions with NASA partnerships enabling renewed American activities on the lunar surface.

4.2 Space Technologies

NASA's wish list for the future of space technology is being written by a NASA specialist as follows, which defines a Roadmap for **Space Technology in different areas in the coming years.**[2] **This Roadmap is updated every few years by NASA as needed**.

"NASA's integrated technology roadmap, which includes both 'pull' and 'push' technology strategies, considers a wide range of pathways to advance the nation's current capabilities in space. Fourteen draft Space Technology Area Roadmaps comprise the overall integrated map."

NASA developed the set of draft roadmaps for use by the National Research Council (NRC) as an initial point of departure for mapping NASA's future investments in technology. Through an open process of community engagement, the NRC will gather input, integrate, and prioritize each Space Technology Area Roadmap, providing NASA with strategic guidance and recommendations that inform the technology investment decisions of NASA's space technology activities. Because it is difficult to predict the wide range of future advances possible in these areas, NASA plans to update the integrated technology roadmap on a regular basis. In 2010–2012 NASA developed a set of 14 Technology Roadmaps to guide the development of space technologies. These Roadmaps were presented in the first edition of this book and the 2015 NASA Technology Roadmaps expanded and updated the original 2012 roadmaps, providing extensive details about anticipated NASA mission capabilities and associated technology development needs. NASA believes sharing the roadmaps with the broader community will increase awareness, generate innovative solutions to provide the capabilities for space exploration and scientific discovery, and inspire others to become involved in America's space program. The 2015 NASA Technology Roadmaps are a set of documents that consider a wide range of needed technology candidates and development pathways for a period of 20 years (2015–2035). The 15 Roadmaps are as follows[2]:

Figure 4.1 below shows the communication and navigation roadmap and Figure 4.2 shows the roadmap for the Science Instruments, Observatories, and Sensor Systems.

For each technology area, NASA established an internal team of subject matter experts who could reach out to other experts throughout NASA as needed. In some cases, the teams reached out to governmental agency experts, and to industry experts through fact-finding meetings. Figure 4.3 show that each team created a technology area breakdown structure.

Each of these teams would then do the following:

a. Identify the top technical challenges that, if met, would achieve needed performance.

b. Identify the mission "pull" technologies needed to support the increased capabilities demanded by future planned NASA missions.

c. Identify emerging "push" technologies that could meet NASA's long-term strategic challenges.

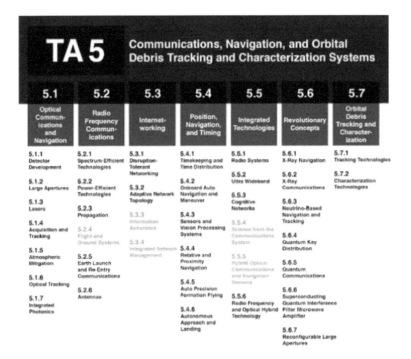

FIGURE 4.1
Technology area breakdown structure technology areas for communications, navigation, and orbital debris characterization systems area strategic roadmap. (TASR.)

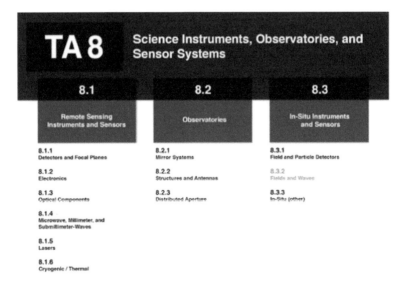

FIGURE 4.2
Technology area breakdown structure (TABS) for science instruments, observatories, and sensor systems.

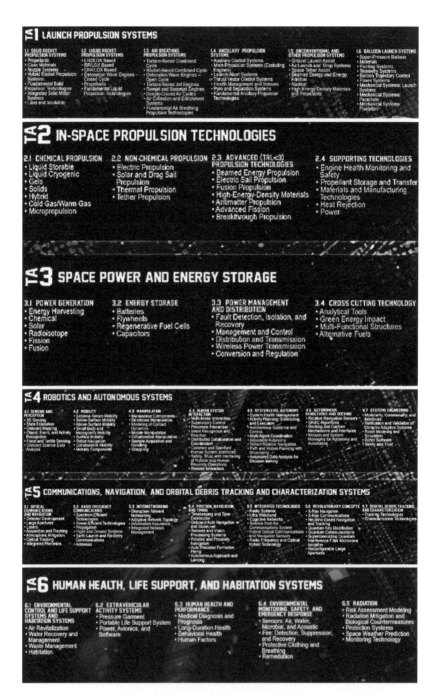

FIGURE 4.3
2015 NASA technology roadmap.

(Continued)

FIGURE 4.3 (CONTINUED)
2015 NASA technology roadmap.

(Continued)

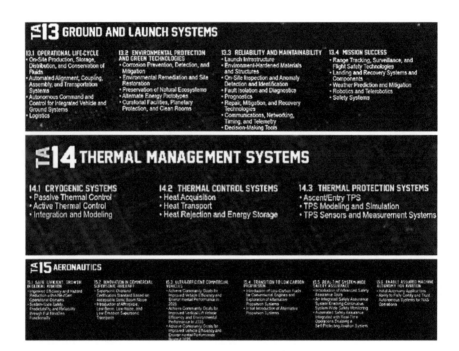

FIGURE 4.3 (CONTINUED)
2015 NASA technology roadmap.

4.3 Non-rocket Space Launch

Rocket-based space launch is very expensive. Numbers such as $10,000 per kilogram is a normal cost for a rocket-based space launch. Table 4.1 shows the heavy lift launch vehicles worldwide.[3] The super-heavy launch vehicles are shown in Table 4.2.

Non-rocket launchers are being introduced in space-related activities which will by some estimates revolutionize the space industry. These non-rocket launchers include the cable accelerator, circle launcher and space keeper, skyhooks, space elevator transport system, space towers, kinetic towers, the gas-tube method, sling rotary method, asteroid employment, electromagnetic accelerator, endo-atmospheric tethers, Sun and magnetic sails, solar wind sail, radioisotope sail, electrostatic space sail, laser beam, Slingatron, kinetic anti-gravitator (repulsitor), Earth–Moon or Earth–Mars non-rocket transport system, multireflective beam propulsion system, electrostatic levitation, and so forth.

To get a feeling for how different in cost the rocket-based versus non-rocket-based launch vehicles is, see Table 4.3 in which these two technologies are compared.

Figures 4.4–4.6 show three of these non-rocket-based vehicles.

TABLE 4.1

Heavy Lift Launch Vehicles

Vehicle	LEO (kg)	GTO (kg)	LEO ($/kg)	GTO ($/kg)
Ares 1	25,600			
Ariane 5	21,000	9,600		20,000
Atlas V	29,420	13,000		
Delta IV heavy	22,950	12,980		
Falcon 9	32,000	19,500	3,273	7,500
Long March 5	25,000	14,000		
Proton	21,600	6,300	4,302	18,359
Rus-M	23,500	7,000		
Shuttle	24,000	3,810	10,416	50,874
Titan IVB	21,682	5,761	11,530	43,395

TABLE 4.2

Super-Heavy Launch Vehicles

Vehicle	LEO (kg)	TLI (kg)
Ares V	160,000	63,000
Energia	100,000/20,000 (GTO)	32,000
N1	75,000	
Saturn V	118,000	47,000
Saturn INT-21	75,000	

TABLE 4.3

Comparison between Rocket-Based and Non-Rocket-Based Launching Vehicles

Method	Estimated Cost ($B)	LEO (kg)	LEO ($/kg)	Capacity	Tech Ready
Rocket		450–29,610	3,200–40,000	200–500	9
Space elevator	6.2–40	18,000+	220–400	2,000	2–4
Hyper skyhook	<1	1,500		30	2
HASTOL		15,000			2
Orbital ring	39		<0.05		2
Launch loop(s)	20	5,000	300	40,000	2+
Launch loop(l)	61	5,000	3	6,000,000	2+
StarTram	60	70,000,000	<100	600,000	2
Ram accelerator			<500		6
Space gun	3	450	500		6
Slingatron		100			2
Spaceplane	23	12,000	3,000		7
Laser propulsion		?	20		3

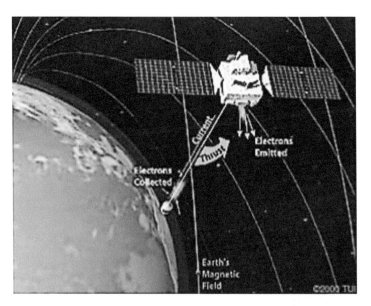

FIGURE 4.4
Principle of electrodynamic tether propulsion.

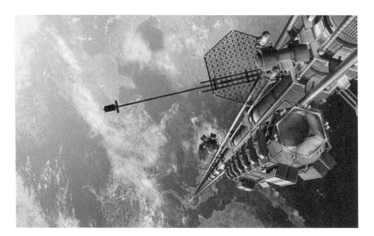

FIGURE 4.5
Space elevator.

4.4 Single Stage-to-Orbit

By definition a single-stage-to-orbit (SSTO) vehicle reaches orbit from the surface of a body without jettisoning hardware, expending only propellants and fluids. The term often refers to reusable vehicles. Although no Earth-launched SSTO launch vehicles have ever been constructed, it

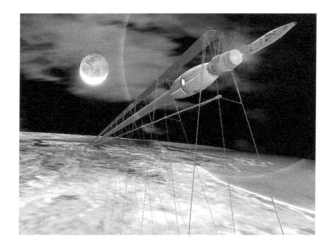

FIGURE 4.6
StarTram. https://en.wikipedia.org/wiki/StarTram#/media/File:Startramgeneration2.jpg.

FIGURE 4.7
Proposed VentureStar.

is an interesting technology that has attracted a lot of attention recently. Several research spacecraft have been designed or constructed, including Skylon, the DC-X, the X-33, and the Roton SSTO. One major obstacle in the successful launch to orbit by SSTO is problems with finding the most efficient propulsion system. SSTO has been achieved from the Moon by the Apollo program's lunar module and several robotic spacecraft of the Soviet Luna program. Having less gravity and almost no atmosphere makes this much easier than from Earth. Figure 4.7 shows the VentureStar which was a proposed SSTO spaceplane, and Figure 4.8 shows VentureStar compared with the space shuttle.

Cost savings (1/10 that of space shuttle) and increased safety makes the VentureStar's engineering and design more attractive than the space shuttle. An airliner-like operation is the ultimate goal for an SSTO vehicle. The development of high-performance hydrogen–oxygen propulsion, pioneered by the General Dynamics Centaur, has encouraged the possibility of SSTO and has made this 40-year-old dream closer to reality.

Table 4.4 summarizes the main SSTO system requirements.[4]

Some unique characteristics of SSTO are as follows:

a. The SSTO will take off straight up, fly to orbit without dropping anything off, and will then land on its tail on a pillar of fire, not gliding but under power.
b. SSTOs will not require strap-on external tanks or boosters.

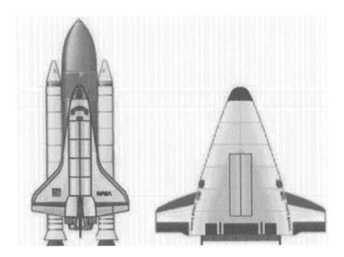

FIGURE 4.8
VentureStar compared with space shuttle.

TABLE 4.4

Key Single-Stage-to-Orbit (SSTO) System Requirements

- Rapid, low-cost turnaround (low-operating costs using only 350 man days)
- Intact abort any time during flight (save the payload and/or crew)
- Medium payloads deployed and/or retrieved (10,000 lb to low Earth orbit)
- Manned and/or unmanned operation (automated flight plus inherent reliability and safety for crews)
- Rocket propulsion as the prime mover (avoid dependency on NASP air breathing technology)

c. SSTOs will not need a long runway, a huge vehicle assembly building, or a mission control building. It will need only a 200-foot diameter concrete pad, a maintenance hangar, and a hydrogen/oxygen propellant facility.

d. SSTOs will navigate using satellite signals from the existing Global Positioning System.

Once fully operational a reusable SSTO will be as safe as flying on a typical commercial airliner.

Unlike the space shuttle, SSTOs will burn only hydrogen and oxygen. Their exhaust will consist primarily of hot but pure water vapor.

An anticipated turn-around will be about 1 day.

4.5 Solar Power Satellites

A solar power satellite (SPS) beams down energy to a reception area on Earth in the form of microwaves. The SPS system requires a global undertaking by all stakeholders from private to public enterprises.

Retired president of India, Dr. Abdul Kalam, spoke publicly on the prospects for space solar power while addressing a symposium on "The Future of Space Exploration" organized by Boston University. Referring to the population growth worldwide in the next 40 years (Figure 4.9), Kalam said, "What better vision can there be for the future of space exploration than participating in a global mission for perennial supply of renewable energy from space?"

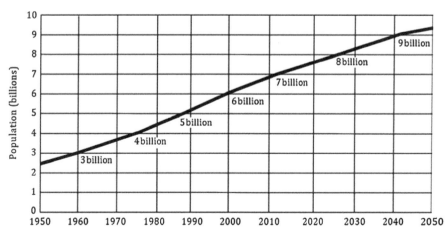

FIGURE 4.9
World population: 1950–2050 (U.S. Census Bureau, International Data Base, 2008 First Update).

Kalam believed that interdisciplinary research on space will enable new innovations in science and engineering. In his address at Boston University he also said, "Civilization will run out of fossil fuels in this century. Solar energy is clean and inexhaustible. However, solar flux on Earth is available for just 6 to 8 hours every day whereas incident radiation on a space solar power station would be 24 hours every day."

The Japanese developed in the System Definition Studies of the late 1970s and conducted some limited testing and proposed a low orbit 10 megawatt demonstration satellite. As one can imagine, the overwhelming initial cost of development and deployment of such a system has remained the primary obstacle. For solar power satellites to become economically feasible there is need for a low-cost space transport, and for SPS to be successful, there is a need for an organized industry and a viable approach for SPS implementation.

SPSs are to be placed in low Earth orbit from which they will transmit the collected solar energy back to Earth by microwave transmission. Such a huge global undertaking needs the cooperation of different nations and an organization that could manage and monitor it worldwide. International involvement of UN member countries who are also members of International Telecommunications Union (ITU) is mandatory for coordinating global treaties and agreements, frequency assignments, satellite orbital locations, space traffic control, and other activities to prevent any potential radio interference into their domestic communication systems and services. To make this global project possible, an international organization such as the ITU is needed in which member countries elect the officials of the organization observing equitable geographical presentations and also based on a rotational system in which every member country in every region could participate in decision-making policies that would ultimately impact their nation. A very careful blend of the private sector should also be considered in the design of the system and the delivery of the service without any power in the decision-making processes. The various elements of the project involving government and industry partnerships must be clearly defined. The design, manufacturing, launching, operating, and maintaining such a global system of SPSs in space orbit would with no doubt involve sophisticated engineering technologies and require great human effort and ingenuity to meet all the challenges that are involved in such a project. It has been shown through recent history that if countries put their minds together, nothing that is imaginable and practical from an engineering point of view would be impossible to achieve. This project is no exception to this rule. The distribution of solar energy worldwide using a space-based platform such as the one described here would provide an alternative source of energy to any point on Earth and would impact the global economy.

As we can see from Figure 4.10, the SPS concept is more than 40 years old, and it has to be considered a viable source of energy by several countries before it can become a reality.

Figure 4.11 illustrates the different components of a possible SPS project.

•	1968	Peter Glaser proposed the concept
•	1972	NASA/Lewis evaluated the concept
•	1973	**Glaser patented the concept**
•	1976	NASA demonstrates microwave power transmission
•	1977	Department of Energy (DOE)/NASA initiated Concept Development and Evaluation Program (CDEP)
•	1978	DOE/NASA held first programmatic review
•	1979	DOE/NASA held second programmatic review
•	1980	DOE held SPS Symposium and third program review
•	1980	DOE conducted peer reviews and published CDEP environmental, societal, and comparative results
•	1980	DOE/NASA terminate further SPS specific research activities
•	1981	National Academy of Sciences published results of SPS critique
•	1981	Office of Technical Assessment published SPS assessment
•	1990s	Renewed interest in concept

FIGURE 4.10
SPS overall design concept (Billion Year Plan), http://billionyearplan. blogspot.com/2010/08/how-far-weve-come-spacebased-solar.html.

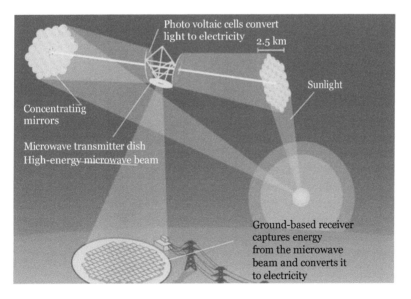

FIGURE 4.11
History of the SPS concept.[6]

4.6 Satellites and Cyberattacks

Internet of Things (IoT) will present enormous challenges for people who offer satellite communication products and services because each connected device from refrigerators to tractors offers hackers a point of entry into the network and a way to target other elements of the network.

At the rate that IoT is propagating through almost every aspect of our lives, in the next few years, the number of connected devices in homes, offices, farms, and transportation hubs is expected to skyrocket. IoT will in effect simplify life by offering a way for people to remotely control appliances and maintain broadband connections while traveling in cars, trains, ships, and aircraft. The beauty of the IoT is that not only you can control your devices remotely but also in the near future the technology will allow you to interconnect these different devices and program them to function as one unit in order to optimize the available resources.

Intelsat's Network Operations Center mages and monitors its customers' satellite traffic. Credit: Intelsat 17

"The more we integrate, the more vulnerabilities we see," Donna Bethea-Murphy, global regulatory senior vice president for telecommunications satellite operator at Inmarsat, said March 8 during a cybersecurity panel at the Satellite 2017 Conference.

That threat was illustrated in October when hackers used connected cameras and digital video recorders to take down Amazon, Twitter, Netflix, and other popular Internet sites. "That is the risk," Duggal told *SpaceNews*. "Your connected fridge could attack you."

http://spacenews.com/whos-keeping-satellites-safe-from-cyberattacks/.

To fight back, satellite equipment and service providers are taking a variety of steps, including updating their own security procedures, carefully selecting partners and sharing threat information.

"We have to stay vigilant," said David Henning, Hughes Network Systems director of network security. "There are always going to be a variety of groups out there that are going to try to go after us."

For satellite and teleport operators, those groups include individual hackers trying to disrupt communications as well as sophisticated nation states.

"We are not only subject to the classic commercial-threat actors, but because we carry communications that are mission critical, we are targeted by nation-state actors," said Andrew Tomaszewski, chief information officer and chief information security officer for VT iDirect, a VSAT equipment manufacturer based in Herndon, Virginia.

For iDirect, defense begins with testing and repairing the products, it sells rather than relying on customers to perform that work. iDirect also hires third-party experts to evaluate its software code and the firm runs an incident response program.

"If our customer or partner reports incidents or security issues, we have a response protocol that's built into the company," Tomaszewski said during the panel. "Everyone knows what to do when one of those calls comes in."

Vigorous incident response protocols are becoming more widespread. ITC Global, which offers satellite communications for remote energy, mining, and maritime customers, is owned by Panasonic. The company operates a security operations center around the clock with a team "purely focused on security," Chris Hill, ITC Global chief technology executive, said during the panel.

Increasingly, satellite communications firms also are taking a close look at the security procedures employed by their partners and customers because any weak link can compromise the network.

Intelsat, for example, pays cybersecurity firms to assess the VSAT equipment its customers use to gain access to satellite links. "We pay for that ourselves," Duggal said. "There is an automatic, default expectation that security is built into VSAT platforms. It is not. When we go to market as an industry there are a lot of moving parts."

Overall, Intelsat spends about 5% of its technology budget on information security. "If you address security up front and make it a part of your engineering cycle, it reduces the cost dramatically," Duggal said. "For those that complain security is expensive, it's because they are addressing it after the fact. It's expensive to reengineer systems, solutions and software."

Companies also evaluate their component suppliers. Hughes Network Systems, for example, turned down business in the past when a customer explicitly requested equipment from a manufacturer the U.S. government cited as a potential threat.

Customers and partners respond in dramatically different ways to conversations about cybersecurity.

"With NATO, it's the first part of the discussion and all the way through the discussion," Simon Gray, humanitarian affairs vice president for satellite fleet operator Eutelsat, said during the panel. "If you are dealing with any military customer, that is one of their first questions."

"The same is true for customers who oversee critical national infrastructure, like pipelines," said Dave Rehbehn, Hughes Network Systems international division vice president. "They take security to an extreme," he told *SpaceNews*.

Other customers pay little attention. "Some customers are set up to absorb information and act on it," Hill said. "Other customers' eyes glaze over and they say, 'I'm not sure what you are talking about.' "In those cases, the service provider needs to help them understand the problem and what is at stake," Hill added.

Those discussions are an important element of cybersecurity. "As cyber threats proliferate across systems and networks, our industry is responding to ensure stability," said Bethea-Murphy, who led the Satellite Industries Association and Global VSAT Forum's joint cybersecurity working group. "We do this through communications. We have to talk to each other at every level, whether it is directly with customers, with manufacturers, with government agencies."

4.6.1 Staying a Step Ahead of a Cyberattack

It is clear from the above that companies are paying more attention to the cyber security and take measures to eliminate any potential risk of being hacked by invaders which if successful will ultimately impact the end user and jeopardize the trust of the customers. As a result, satellite operators are adopting safeguards to ensure the same features that enable communications satellites to respond quickly to changing demand and prevent spacecraft from cyberattack.

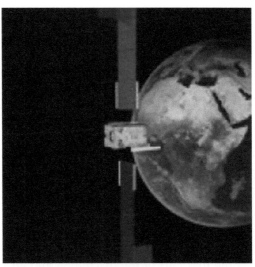

Skynet 5 satellite. Credit: Airbus.

Britain's Skynet 5 military satellite telecommunications system, managed by Airbus Defense and Space, offer customers the ability to change the shape and power levels of communication beams. Intelsat may incorporate such beam-shaping technology on future EpicNG satellites.

"You can change the shape of the beam whenever you want to from the ground," said Mark Daniels, Intelsat General Corp.'s vice president for new technologies and services. "You can start with a beam shaped for a particular region but if demand changes or there are problems with jamming, you can reconfigure the beam to increase the coverage area or notch out an area to avoid jamming."

The same flexibility that helps operators mitigate jamming problems, could pose a hazard if an attacker took control of a satellite. "Anytime you have something up there with that much flexibility, you need more cybersecurity around it because someone might take advantage of that flexibility," Daniels said.[8]

4.7 Photonics in Space

The enabling potential of photonics in space engineering has notably grown since the 1960s, when the only photonic devices on satellites were the solar cells. In recent years, photonic components and sub-systems have been crucial for many functions on boarding a spacecraft. Functions such as data handling, attitude and orbit control, as well as strain/thermal mapping use photonics technology.

Currently, many payloads for both Earth observation and scientific missions include a wide variety of optical and opto-electronic components, such as lasers, detectors, modulators, lenses, gratings, mirrors, and so on.[9]

Photonics can play a key role in nearly all satellite sub-systems (shown in Figure 4.12) due to its intrinsic advantages with respect to conventional technologies. Opto-electronic gyroscopes have routinely been included in attitude and orbit control systems for several decades, and some important advantages of optical fibers in the implementation of on-board data buses have been proved in several space missions, starting from the 1990s. Since their first utilization in space, in 1958, the solar cells have been included in the power supply sub-system of all satellites.

Mapping of strain and temperature in some critical sections of the spacecraft can benefit from fiber Bragg grating technology, whose space applicability has already been proved. Optical wireless links to transfer data from one satellite to another or from a satellite to a ground station have also been successfully demonstrated in some recent space missions.

In terms of the principles of photonics, it could be said that photonics is the generation, detection, and manipulation (amplification, modulation, processing, switching, steering) of photons. It refers to guided wave technologies in either optical fibers or waveguides. Photonics are expected to make an impact in future Spacecraft engineering by replacing or enhancing conventional electrical approaches in the fields of digital and RF telecom

Sub-system	Basic function
Mechanical structure	Accommodate all other sub-systems
Power supply	Generate and efficiently distribute electrical energy within the spacecraft
Thermal sub-system	Keep the temperature of spacecraft components within appropriate intervals
Attitude control	Real-time monitor and control the attitude of the spacecraft in space
Communications	Transmit/receive data (telemetry data, commands, and payload data) to/from the ground stations
Data processing and handling	Process and handle data on board the spacecraft
Propulsion	Change the orbit of the spacecraft

FIGURE 4.12
Satellite sub-systems.

payloads, sensors, micro LIDARs, and spectrometers by reducing the size, weight, power, or performance of the systems they replace.[10]

Some of the applications of photonics in space systems could be summarized as follows[10]:

4.7.1 Flexible RF Payload

Microwave photonics are being developed to offer new functionality and performance to RF payloads. At the core of these payloads are photonics to generate stable local oscillators, perform optical down conversion, and manipulate the optical microwave signals by routing, beamforming or filtering them in the optical domain. The main components of such a photonic system include lasers, optical amplifiers, WDM components, optical modulators, optical switches, and photonic integrated circuits for optical beamforming, switching, and filtering.

4.7.2 Optical Interconnects for Digital Payloads

New optical transceiver hardware is under development to allow high-speed (25 Gbps) optical digital links to interconnect future digital processors. In addition, European Space Agency (ESA) is looking at how photonics can be used to improve the performance of analogue to digital conversion. The use of low loss optical fibers also allows high-density optical interconnections making use of high density, lightweight optical fiber harness, and flex foils for the distribution of digital signals within these payloads, giving increased freedom to the system designers. The optoelectronics section is participating in the full development cycle of these new payloads, from component specification and evaluation to system level breadboarding.

4.7.3 Photonic Integrated Circuits

Photonic integrated circuits are the chip scale integration of multiple optical elements or components, which enable complex functions analogous to the electrical integrated chips. As these chips increase in complexity and functionality they are finding new space applications; micro spectrometers, integrated solid state gyro, laser beam steering, complex optical modulation/demodulation, optical switching, optical beam forming, packet processing. The main advantage of this approach is clearly to target a massive size and weight advantage, but also the advantage of potential cost reduction (manufacturing, assembly, and qualification).

4.7.4 Inter-Satellite Digital Communications

In order to reduce the time delay of a double-hop satellite links, the signal received by a satellite is sent to the adjacent satellite from which it is downlinked to the receding station. Inter-Satellite links (ISL) have been in use for many years using the 40–60 GHz frequency band. With the new technology, Space Fiber is used which is a multi-Gbits/s, on-board network technology for spaceflight applications, which runs over electrical or fiber-optic cables. The Optoelectronics section together with European partners has been developing 10 Gbps optical transceivers to meet the needs of future satellites, and have been involved in component testing and evaluation to IODs at system level.

4.7.5 Fiber-Optic Sensing for Satellite Platforms

Fiber-optic sensing is a new tool in spacecraft engineering, which permits the measurement of: temperature, strain, acceleration, and rotation by modulation of some parameter of light propagating inside a fiber. The optoelectronics section has been exploiting these techniques to understand how this fiber technology can be used in future space missions. The technologies under investigation include Fiber Bragg Gratings (FBGs), Photonic Band Gap sensors, In-fiber interferometers (FOG, accelerometers), as well as distributed-sensing approaches using the natural scattering mechanisms of the fiber itself which permit 1,000s of measurement points per meter of the fiber.

4.7.6 Photonics for Launchers

Optopyrotechnics is a new approach to the detonation of pyrotechnics using short pulses of a laser output to ignite the pyro material. This technology has been baselined for the next European launcher, the Ariane 6. The opto-electronics section together with European industry has been leading the development of key components of this system, from the laser to the safety features such as the optical safety barrier. Other photonic technologies have been studied for use in future launchers include, laser ignition, fiber-optic sensing, optical communications, and optical wireless.

Spacecraft prime contractors are slowly recognizing the benefits of fiber optics for on-board data handling, benefits include:

- Total Elimination of EMI and RFI
- Significant Reduction in Size, Weight, and Power
- Greater On-orbit and I&T Flexibility
- Significantly Reduced I&T Costs

4.8 Questions

1. According to NASA report *Voyages: "Charting the Course for Sustainable Human Space,"* what are the vision of the next steps out into the solar system?

2. NASA developed the set of roadmaps in 2012 for use by the NRC, which were later modified in 2015, name the areas outlined in this roadmap.

3. For each technology area, NASA established an internal team of subject matter experts who could reach out to other experts throughout NASA as needed. What are the functions of these teams?

4. Briefly explain the potentials of cyberattacks on satellites.

5. What did Britain's Skynet 5 military satellite telecommunications system, managed by Airbus Defense and Space, offered to its customers?

6. Name the key roles that photonics can play in satellite sub-systems.

7. Name some of the applications of photonics in space systems.

8. What are the benefits of Fiber Optics for On-board Data Handling?

9. Explain the Inter-Satellite Link and its impact on the quality of satellite communications.

10. Explain SPS overall design concept.

References

1. National Aeronautics and Space Administration Langley Research Center. www.nasa.gov/centers/langley.
2. www.nasa.gov; www.nasa.gov/offices/oct/home/roadmaps/index.html.
3. Non-rocket space launch, www.thefulfillment.org/docs/NonRocketSL.ppt.
4. Heald, D.A., and T.L. Kessler. Single stage to orbit vertical takeoff and landing concept technology challenges engineering manager. Advanced Space Concepts SSTO Chief Engineer General Dynamics Space Systems Division, San Diego, CA, www.spacefuture.com/archive/single_stage_to_orbit_vertical_takoff_and_laning_concept_technology_challenges.shtml.
5. Koomanoff, F.A., and C.E. Bloomquist. 1993. *Solar Power Satellites*. New York: Ellis Horwood, p. 26. http://fti.neep.wisc.edu/neep602/LEC32/IMAGES/fig5.GIF, http://fti.neep.wisc.edu/neep602/lecture32.html.
6. www.youtube.com/watch?v=xGLCs3nVXhA.
7. Debra Werner. Retrieved on April 19, 2017. http://spacenews.com/whos-keeping-satellites-safe-from-cyberattacks/.
8. http://spacenews.com/whos-keeping-satellites-safe-from-cyberatacks/#sthash.LKtpAN3.dpuf.
9. www.worldscientific.com/worldscibooks/10.1142/9817.
10. www.esa.int/Our_Activities/Space_Engineering_Technology/Space_Optoletronics/Photonics.

5

Information, Communication, and Space Technologies

5.1 Basic Definitions and Principles of Information and Communication Technology

Information and communication technology (ICT) by definition is an umbrella term that includes any communication device or application, encompassing radio, television, cellular phones, computer and network hardware and software, satellite systems, as well as the various services and applications associated with them. Such applications span from videoconferencing and distance learning to telemedicine, e-government, e-commerce, e-education, e-health, and e-environment. ICT applications can facilitate the achievement of reducing poverty and improving health and environmental conditions in developing countries.

One could also define ICT as a technology that covers any product that will store, retrieve, manipulate, transmit, or receive information electronically in a digital form.

According to the Information and Communication Technology Sector Strategy Paper of the World Bank Group in April 2002,[1] ICT is defined as the technology that consists of hardware, software, networks, and media for collection, storage, processing, transmission, and presentation of information in forms of voice, data, text, or image.

Realizing the ever-widening gap between the developed and developing countries, the International Telecommunications Union (ITU) decided to take advantage of ICT as a major tool in order to close the gap between them. It is considered as the missing link that once integrated into the policies of developing countries, will positively impact the socioeconomical well-being of those countries. With sufficient investment and the right approach and implementation processes, ICT applications and tools can result in productivity and quality improvements.

ICTs provide many opportunities such as making learning more interesting, especially for hard-to-understand issues; bridging distances (e.g., using e-mails, phone, videoconferencing, and so forth); breaking literacy barriers in communication (e.g., using video and radio); and sharing research and useful information (e.g., using the Internet). ICT can also provide access information on jobs and internships, create new employment opportunities, enhance interaction with peers over long distances, create entertainment opportunities (games, music, video), and provide more realistic information on life elsewhere. ICT covers such a wide spectrum of technologies and applications that it is wise to allocate part of the curriculum to discussing what ICT is and how it could be effectively utilized in schoolwork and beyond graduation in work environments.

There is in some cases a resistance in adopting ICT within educational practices. The main factor that prevents some teachers from making full use of ICT can be broadly categorized into the following areas. At the teacher level, the factors span from the lack of teachers' ICT skills, lack of teachers' confidence in adopting ICT as a tool for teaching, lack of pedagogical teacher training, lack of follow-up of new ICT skills, and finally the lack of differentiated training programs. At the school level, the factors are the absence of ICT infrastructure, old or poorly maintained hardware, lack of suitable educational software, and limited access to ICT. Other school-related factors include limited project-related experience and lack of ICT mainstreaming into the school's strategy. Finally, at the system level, factors such as the rigid structure of the traditional education system, traditional assessment, restrictive curricula, and restricted organizational structure are to be considered.

Figure 5.1 shows ICT application categories.[2]

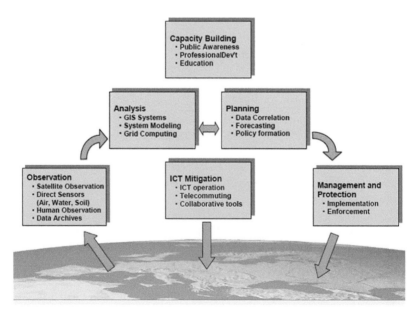

FIGURE 5.1
ICT application categories (ITU).

One of these areas which are highly dependent on satellites technology is the Environmental Capacity Building. Increasing public awareness of environmental issues, their impact on humans and nature, the long-term impact on generations to come, and integrating environmental issues into formal education would build the capacity for addressing such environmental challenges that we are facing today. Remote sensing, Geographic Information System (GIS), and other means of data collection, analysis, and interpretation should be not only part of the curriculum at different levels of education but also an essential part of the overall policy that reaches beyond educational circles.

Advances in the use of remote-sensing technologies over the last four decades since its inception, as was explained in more details in Chapter 3, have allowed detailed observation and monitoring of the Earth and its atmosphere that have benefited mankind. The first commercial Earth Resources Satellite, LANDSAT-1, which was launched in 1972, made images available worldwide with a rather good resolution at the time. The major advantages of remote-sensing satellites are its larger geographical coverage and its repetitive nature of data collection. The repeated coverage of the same point every 18 days from LANDSAT generations 1–3, and 16 days thereafter, provides invaluable information in a variety of application areas. The global observation system of the World Meteorological Organization (WMO)[3] is a satellite-based platform for observation and communication which provides information on barometric pressure, water temperature, and wave action to the users of the system. In such a system in order to supplement and revalidate the satellite data, land-based sensors using radio telemetry relay data are acquired from surface stations. Commercial aircraft, specialized weather observation aircraft, weather balloons, and ships are also used as complementary platforms. Figure 5.2 displays the Global Observing System with the above-mentioned platform shown.

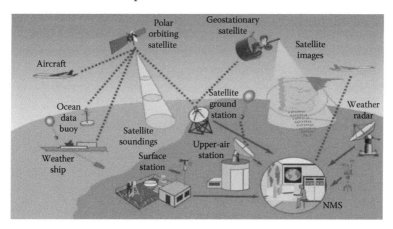

FIGURE 5.2
Global observing system (World Meteorological Association, www.wmo.int/pages/prog/www/OSY/GOS.html[4]).

5.2 Emerging Applications of ICT

Incredible changes and exponential growth have been observed in ICTs over the past decade. Moore's Law continues to hold true as hardware gets smarter, faster, and smaller, and Nielsen's Law also continues to hold as bandwidth continues to increase and become more affordable. In this section, we will take a brief look at some of the new and emerging technologies, applications, services, and access that will continue to push us forward.

- **Landline Connectivity**
  Most of us still rely on some sort of a "hard" connection in our homes. For some of us it is a fiber optic connection run right to our home, for others it is still a copper connection. Let's take a look at some of the different technologies involved.

- **Digital Subscriber Line**
  Digital Subscriber Line (DSL) uses the existing copper pair or local loop to provide broadband data services along with Plain Old Telephone Service (POTS) for voice. There are lots of variations of DSL with each variation, in acronym, ends with the DSL suffix. The prefix is typically different so "x"DSL is typically used to refer to all variations of digital subscriber line services. Asymmetric Digital Subscriber Line (ADSL) is currently the most popular offering from providers across the United States. DSL allows telephone companies to extend the life of copper wire pairs a little longer. DSL splits frequencies with the lower frequencies used for voice and the higher frequencies used for data as illustrated in Figure 5.3.

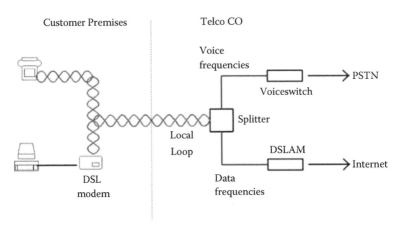

FIGURE 5.3
Typical digital subscriber line (DSL) network configuration.

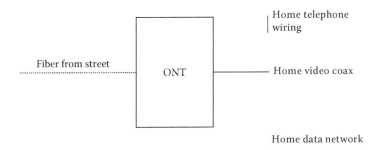

FIGURE 5.4
Fiber to the home.

DSL has distance limitations and the local loop typically needs to be cleaned up of any splices and devices called loading coils. Data bandwidth typically runs between 250 Kbps (Kilo bits per second) and 40 Mbps (Mega bits per second).

- **Fiber to the Home**
  In 2006, in technical competition with the cable companies, Verizon decided to go with a Fiber to the Home (FTTH) implementation—what we know as FiOS. FiOS runs an optical fiber directly to a subscriber's home to deliver voice, video, and data services. Total Verizon investment estimates for the product implementation are anywhere between $18 and $23 billion. Initial Verizon plans were to pass 3 million homes per year starting in 2006 until they reached approximately 60% of their 2006 customer footprint. FTTH technology involves running a piece of fiber directly to a residence and connecting the optical fiber to an Optical Network Terminal (ONT). The ONT converts optical signals to electronic signals as illustrated in Figure 5.4. There is also a replaceable battery as part of the ONT configuration to provide power to the phones in the event of a power failure. Operating at Terahertz frequency band results in theoretical data bandwidths for FTTH exceeding 1,000 Mbps.

- **Fiber to the Node**
  Around the same time Verizon was planning FTTH, AT&T was planning a rollout of Fiber to the Node (FTTN) technology. FTTN involves running fiber into a neighborhood but, instead of running it right up the house like FTTH, FTTN runs fiber out to a pedestal on the side of the road. Electronics in the pedestal convert optical signals to electrical signals with the connection completed to the home over the existing copper wire pair originally used for the telephone. This is shown in Figure 5.5.
  FTTN is essentially a version of DSL called Very-high-bitrate DSL (VDSL or VHDSL). At the house, the copper pair goes into a network interface similar to the FTTH ONT with voice, video, and data splitting off.

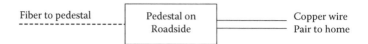

FIGURE 5.5
Fiber to the node.

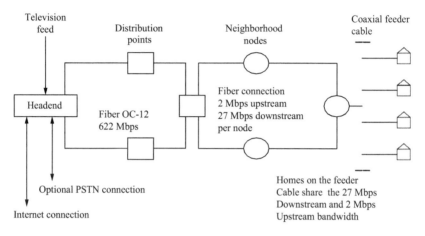

FIGURE 5.6
Typical cable network connection.

- **Cable Modem**

  Cable TV networks were originally constructed for the delivery of downstream traffic only. To provide the necessary upstream bandwidth, cable providers modified their networks for two-way traffic. Today most cable companies that have moved into both the voice and data business have modified their network to support simultaneous upstream and downstream transmission. These TV/data networks are typically Hybrid Fiber Coaxial (HFC) with fiber used for trunking and feeding while coaxial cable is still typically used to bring the lines into a home or business as shown in Figure 5.6.

  A modern cable headend serves 100,000–300,000 customers, and distribution points will serve 10,000–30,000 customers. Neighborhood nodes can serve up to 1,000 customers.

5.3 Wireless Connectivity

These days, most of us are carrying some sort of mobile device with the expectation of continuous connectivity and availability. Most recently, we have been exposed to a lot of 4G advertising from wireless providers.

5.3.1 Earlier Wireless Standards

Before we take a look at 4G technologies, let's take a quick look at some of the legacy technologies that got us to where we are today.

1G—First-generation wireless introduced in the 1980s, uses analog voice modulation at 150 MHz and up.

2G—Second-generation wireless introduced in the 1990s similar to 1G except radio signals on 2G networks are digital; originated as purely circuit switched, later expanded to packet switched.

3G—Third-generation wireless introduced in the 2000s; provides peak data rates for existing services up to 28 Mbit/s downstream and 22 Mbit/s upstream.

Notice about once every 10 years a new-generation wireless technology comes out. These technologies are typically not backward compatible.

5.3.2 Fourth-Generation (4G) Services

4G is short for fourth-generation, and is a successor to third-generation (3G) wireless technologies. 4G includes both LTE (Long-Term Evolution) and WiMAX (Worldwide Interoperability for Microwave Access) and sets peak mobile download speeds of 100 Megabits per second (Mbps) and 1 Gigabit per second (Gbps) for fixed services. An example of a fixed service would be an antenna on top of your house used for wireless access.

You may be wondering if both are considered 4G technologies and both offer the same bandwidths, what is the difference between LTE technology used by providers like Verizon and AT&T and WiMAX used by other companies like Sprint? Is one technology better than the other? Why would one company decide on LTE while another decides on WiMAX to deliver next-generation services? If they are similar, what makes them different?

- LTE is the 4G technology of choice of the larger mobile carriers like Verizon Wireless (launched LTE recently) and AT&T Wireless (scheduled to start LTE rollout this year). These carriers already have LTE spectrum bands and the money to buy more spectrum bands. They will also tell you that LTE more easily supports backward compatibility with earlier cellular technologies. LTE uses Frequency Division Duplex (FDD) spectrum bands.
- WiMAX is the choice of carriers with Time Division Duplex (TDD) spectrum bands (launched by Sprint in 2008) and also makes sense for greenfield situations where backward compatibility is not needed.

Therefore, LTE uses FDD spectrum bands and WiMAX uses TDD spectrum bands what's the difference? Here is a quick explanation from three-g.net:

> Frequency Division Duplex (FDD) and Time Division Duplex (TDD) are the two most prevalent duplexing schemes used in broadband wireless networks. TDD is the more efficient scheme, however, since it does not waste bandwidth. FDD, which historically has been used in voice-only applications, supports two-way radio communication by using two distinct radio channels. Alternatively, TDD uses a single frequency to transmit signals in both the downstream and upstream directions.

Basically, FDD (LTE) uses two channels and TDD (WiMAX) uses one channel for two-way communications.

Which technology will dominate? It looks like LTE will dominate in the United States, but there are already issues with expensive and crowded spectrum bands. There are also a lot more TDD spectrum bands available than FDD spectrum bands, and TDD spectrum bands are cheaper.

Now, if LTE is preferred by the larger carriers in the United States and spectrum bands are in short supply, would it not make sense to try and develop a version of LTE that could use TDD spectrum bands? That is what an emerging technology called TD-LTE does—it uses TDD spectrum for LTE transmission. Does it work? It sure looks like it. In July 2010, Ericsson and China Mobile demonstrated an end-to-end TD-LTE solution that achieved a single-user peak downlink rate of 110 Mbps, and in December 2010, China Mobile announced it had finally received approval from regulators and will start large-scale testing of domestically developed TD-LTE technology.

5.3.3 5G Technology

The 5G technology will provide a wider spectral bandwidth per frequency channel. Some of the characteristics of 5G are as follows:

- High increased peak bit rate
- Larger data volume per unit area
- High capacity to allow more devices connectivity concurrently and instantaneously
- Lower battery consumption
- Better connectivity irrespective of the geographic region, in which you are
- Larger number of supporting devices
- Lower cost of infrastructural development
- Higher reliability of the communications.

The following table shows comparison of 1G–5G Technologies.

Technology/Features	1G	2G/2.5G	3G	4G	5G
Start/Deployment	1970/1984	1980/1999	1990/2002	2000/2010	2010/2015
Data Bandwidth	2 kbps	14.4–64 kbps	2 Mbps	200 Mbps to 1 Gbps for low mobility	1 Gbps and higher
Standards	AMPS	2G: TDMA, CDMA, GSM 2.5G: GPRS, EDGE, 1xRTT	WCDMA, CDMA-2000	Single unified standard	Single unified standard
Technology	Analog cellular technology	Digital cellular technology	Broad bandwidth CDMA, IP technology	Unified IP and seamless combination of broadband, LAN/WAN/ PAN and WLAN	Unified IP and seamless combination of broadband, LAN/ WAN/PAN/WLAN and wwww
Service	Mobile telephony (voice)	2G: Digital voice, short messaging 2.5G: Higher capacity packetized data	Integrated high quality audio, video and data	Dynamic information access, wearable devices	Dynamic information access, wearable devices with AI capabilities
Multiplexing	FDMA	TDMA, CDMA	CDMA	CDMA	CDMA
Switching	Circuit	2G: Circuit 2.5G: Circuit for access network & air interface; Packet for core network and data	Packet except circuit for air interface	All packet	All packet
Core Network	PSTN	PSTN	Packet network	Internet	Internet
Handoff	Horizontal	Horizontal	Horizontal	Horizontal and Vertical	Horizontal and Vertical

Source: http://ids.nic.in/TnI_Jces_May%202012/PDF1/pdf/1.5g_tech.pdf.

5.4 Smart Phones and Tablets

Most of us are carrying around some kind of "smart" connected mobile device that we can use as a phone, for data, to listen to audio, and to watch videos. These devices are redefining the way we work, entertain ourselves, and stay connected with family and friends. The ability for developers to create custom applications for these devices has spawned a whole new industry.

The Pew Research Center's Internet and American Life Project released an interesting report in September 2010 titled "The Rise of the Apps Culture." The report takes a look at cell phone use in the United States and how it has increased and changed over the past decade. The project surveyed 1917 cell phone users. Here is a list of some of the key findings from the report:

- Eight in ten adults today (82%) are cell phone users.
- About one-quarter of adults (23%) now live in a household that has a cell phone but no landline phone.
- Eleven percent of cell phone owners are not sure if their phone is equipped with apps.
- Thirty-five percent of adults have cell phones with apps, but only two-thirds of those who have apps actually use them.
  - Twenty-nine percent have downloaded an app to their phone.
  - Thirty-eight percent have purchased a phone with preloaded apps.
- App use still ranks relatively low when compared with other uses of cell phones.
- One in ten adult cell phone users (10%) had downloaded an app in the past week; 20% of cell phone users under age 30 download apps this frequently.
- One in eight adult cell phone users (13%) has paid to download an app.
- Among cell phone users with apps, the average adult has 18 apps on his or her phone.

Application development opportunities will continue for entrepreneurs as smaller, faster, and better connected devices continue to be developed.

5.5 Future Technology

Inexpensive devices, ubiquitous connectivity, and affordability will continue to drive ICTs into the future. Much of the push will be in the wireless and mobility area. In February 2010, U.S. President Barack Obama released his plan that would enable businesses to grow faster; students to learn more; and public safety officials to access state-of-the-art, secure, nationwide, and interoperable mobile communications. Specifically the plan will:

- Nearly double wireless spectrum available for mobile broadband.
- Set a goal of 98% of Americans with access to 4G high-speed wireless.
- Set up a wireless innovation (WIN) fund to help drive innovation.
- Develop and deploy a nationwide, interoperable wireless network for public safety.

These are all important goals as we move forward.

Communications satellites have been instrumental in providing the necessary last mile connectivity in order to extend the reach of ICT services to remote and isolated regions worldwide. The United Nations in its March 2009 meeting in Bangkok regarding the information, communication, and space technologies for disaster risk reduction said:

> Because of their ability to provide rapid broadband communication deployment, satellites can form the backbone of a communications system during an emergency. Satellite broadband communication has been recognized as one of the most useful means of supporting disaster response. Satellite communication-based services, providing large bandwidth connectivity, could be rapidly and easily redeployed to other locations when and where needed. IP-based platforms support voice, data and video communications requirements, and a network of relevant nodes within a disaster management system would ensure the flow of information in a timely manner.[5]

The proposed strategy for the next phase of the Regional Space Applications Program for Sustainable Development by the United Nations in February 2007 is shown in Figure 5.7.

One good example of the application of ICT and space technology in disaster management is the process adopted by the Pacific in the Pacific ICT Ministerial Forum in February 2009.[6]

Figure 5.8 shows the natural disaster damages in Asia and the Pacific during 1975–2005 periods. The Sentinel Asia system, according to the definition by the Asia Pacific community, is the Internet-based information sharing system that makes use of Earth observation satellite data for disaster management in the Asia-Pacific region, including 51 organizations from 20 countries and 8 international organizations. Figure 5.9 demonstrates the design of the Sentinel Asia Framework. As we can see in this figure, the information collected from the affected areas is sent to an Earth Observing Satellite. This information gets transmitted to a communication satellite, and from there it is sent to the central Disaster Management Organization for utilization and transfer to end users.

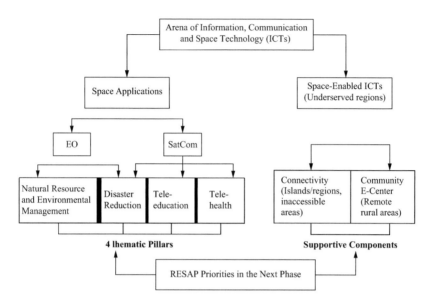

FIGURE 5.7
The proposed strategy for the next phase of the Regional Space Applications Program for Sustainable Development by the United Nations in February 2007. (Economic and Social Commission for Asia and the Pacific, Third Ministerial Conference on Space Applications for Sustainable Development in Asia and the Pacific, Kuala Lumpur Preparatory Meeting of Senior Officials, February 5–7, 2007.[7])

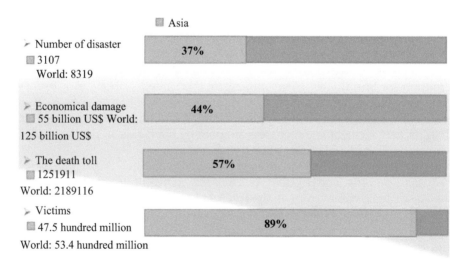

FIGURE 5.8
Natural disaster damages in Asia and the Pacific. (Sentinel Asia: Disaster Management Using ICT and Pace Technology in the Pacific, Pacific ICT Ministerial Forum, February 18, 2009.[8])

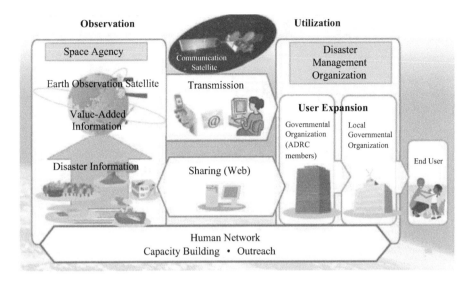

FIGURE 5.9
The design of the Sentinel Asia Framework. (Sentinel Asia: Disaster Management Using ICT and Pace Technology in the Pacific, Pacific ICT Ministerial Forum, February 18, 2009.[8])

Figure 5.10 shows the emergency observation flow. In this picture, we can see that once the disaster occurs, the emergency observation request is generated and sent to the Asia Disaster Reduction Center and up to the satellites while at the same time the camera images are posted on the Sentinel Asia Website. All of this disaster information is then sent to the Disaster Management Agencies in Asian countries.

This example demonstrates how the ICT and satellite systems can work hand in hand to manage a natural disaster occurring in a given region of the world.

ICT together with space technology has brought some revolution to the economies of nations as well. Seidu Mohammed, the director of NASRDA (National Space Research and Development Agency), at the fourth meeting of the Governing Council Board of (ARCSSTE-E) said:

> The vast and unlimited benefits of space science and technology had extended to the fields of communication, sustainable energy, education resources and environmental management, food security, defense and crime control and industrial growth.[9]

Speaking at the same event, the Director/Chief Executive Officer of ARCSSTE-E, Dr. Joseph Akinyede, said:

> space science and technology had become an agent of change impacting on the lives of people and nations around the world.[10]

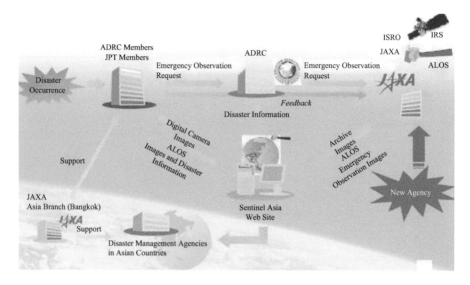

FIGURE 5.10
Emergency observation flow. (Sentinel Asia: Disaster Management Using ICT and Pace Technology in the Pacific, Pacific ICT Ministerial Forum, February 18, 2009.[8])

5.6 Questions

1. Define ICT.
2. What are some cases of resistance in adopting ICT within The educational practices?
3. Name some of the ICT application categories according to ITU?
4. Explain the role that satellite technology plays in ICT as one of its major application area.
5. Name and briefly explain some of the emerging applications ICT.
6. Name the wireless standards and briefly explain each one.

References

1. ICT and MDGs. A World Bank Perspective, World Bank Group, December 2003.
2. www.itu.int/ITU-D/cyb/app/docs/itu-icts-for-e-environment.pdf.
3. World Meteorological Organization (WMO). World Weather Watch Programs, About the Global Observing System (GOS), www.wmo.ch/pages/prog/www/OSY/GOS-purpose.html.

4. World Meteorological Organization (WMO). Global Observing System (GOS), http://www.wmo.int/pages/prog/www/OSY/GOS.html.
5. United Nations Economic and Social Commission for Asia and the Pacific. Committee on Disaster Risk Reduction, E/ESCAP/CDR15, page 13, paragraph 49, Bangkok, March 25–27, 2009.
6. Sentinel Asia: Disaster management using ICT and pace technology in the Pacific, Pacific ICT Ministerial Forum, February 18, 2009.
7. Economic and Social Commission for Asia and the Pacific. Third Ministerial Conference on Space Applications for Sustainable Development in Asia and the Pacific, Kuala Lumpur Preparatory Meeting of Senior Officials, February 5–7, 2007.
8. Sentinel Asia: Disaster management using ICT and pace technology in the Pacific, Pacific ICT Ministerial Forum, February 18, 2009.
9. Mohammed, S. March 2011. Fourth Meeting of the Governing Council Board of the African Regional Centre for Space Science and Technology in English (ARCSSTE–E), Abuja, Nigeria.
10. Akinyede, J. March 2011. Fourth Meeting of the Governing Council Board of the African Regional Centre for Space Science and Technology in English (ARCSSTE–E), Abuja, Nigeria.

Appendix A: Scientists and Mathematicians Referred to in This Book

Alhazen-Ibn al-Haytham

Born:	965 CE (354 AH) Basra in present-day Iraq, Buyid Persia Died: circa 1040 CE (430 AH) Cairo, Egypt, Fatimid Caliphate
Residence:	Basra, Cairo
Fields:	Physicist and mathematician
Known for:	*Book of Optics, Doubts Concerning Ptolemy, On the Configuration of the World, The Model of the Motions, Treatise on Light, Treatise on Place*, scientific method, experimental science, experimental physics, experimental psychology, visual perception, analytic geometry, non-Ptolemaic astronomy, celestial mechanics
Influences:	Aristotle, Euclid, Ptolemy, Galen, Muhammad, Banū Mūsā, Thabit ibn Qurra, al-Kindi, Ibn Sahl, al-Qūhī
Influenced:	Omar Khayyám, al-Khazini, Averroes, Roger Bacon, Witelo, Pecham, Farisi, Theodoric, Gersonides, Alfonso, von Peuerbach, Taqi al-Din, Risner, Clavius, Kepler, John Wallis, Saccheri

Nasīr al-Dīn al-Tūsī

Persian Muslim Scholar

Commemorated on an Iranian stamp upon the 700th anniversary of his death.

Title:	Khaje Nasir
Birth:	February 18, 1201 (11 Jamadi al-Ula, 597)
Death:	June 26, 1274 (aged 73) (18 Dhu'l-Hijjah 672)
Ethnicity:	Persian
Region:	Iran
Maddhab:	Twelver Shī'ah school tradition = Avicennism
Main interests:	Islamic theology, Islamic philosophy, astronomy, mathematics, chemistry, biology and medicine, physics, science
Notable ideas:	Evolution, spherical trigonometry, Tusi-couple
Works:	*Rawda-yi Taslīm, Tajrid al-'Aqaid, Akhlaq-i-Nasri, Zij-i ilkhani, al-Risalah al-Asturlabiyah, Al-Tadhkirah fi'ilm al-hay'ah*
Influences:	Avicenna, Fakhr al-Din al-Razi, Mo'ayyeduddin Urdi Influenced: ibn Khaldun, Qutb al-Din al-Shirazi, Ibn al-Shatir, Copernicus

Johannes Kepler

A 1610 portrait of Johannes Kepler by an unknown artist.

Born:	December 27, 1571 Weil der Stadt near Stuttgart, Germany
Died:	November 15, 1630 (aged 58) Regensburg, Bavaria, Germany
Residence:	Württemberg, Styria, Bohemia, Upper Austria
Fields:	Astronomy, astrology, mathematics, and natural philosophy
Institutions:	University of Linz
Alma mater:	University of Tübingen
Known for:	Kepler's laws of planetary motion, Kepler conjecture

Sir Isaac Newton

Godfrey Kneller's 1689 portrait of Isaac Newton

Born:	January 4, 1643 [OS: December 25, 1642] Woolsthorpe-by-Colsterworth, Lincolnshire, England
Died:	March 31, 1727 (aged 84) [OS: March 20, 1726] Kensington, Middlesex, England
Residence:	England
Nationality:	English
Fields:	Physics, mathematics, astronomy, natural philosophy, alchemy, Christian theology
Institutions:	University of Cambridge, Royal Society, Royal Mint
Alma mater:	Trinity College, Cambridge
Academic advisors:	Isaac Barrow, Benjamin Pulleyn

Gottfried Wilhelm Leibniz

Full name:	Gottfried Wilhelm Leibniz
Born:	July 1, 1646 Leipzig, Electorate of Saxony
Died:	November 14, 1716 (aged 70) Hanover, Electorate of Hanover
Era:	17th-century philosophy, 18th-century philosophy
Region:	Western Philosophy
Main interests:	Metaphysics, mathematics, theodicy
Notable ideas:	Infinitesimal calculus, Monadology, Optimism Leibniz formula for pi, Leibniz harmonic triangle, Leibniz formula for determinants, Leibniz integral rule, Principle of sufficient reason, Diagrammatic reasoning, Notation for differentiation, Proof of Fermat's little theorem, Kinetic energy, Entscheidungsproblem AST

Immanuel Kant

Full name:	Immanuel Kant
Born:	April 22, 1724 Königsberg, Prussia (Now Kaliningrad, Russia)
Died:	February 12, 1804 (aged 79) Königsberg, Prussia
Era:	18th-century philosophy
Region:	Western Philosophy
School:	Kantianism, enlightenment philosophy
Interests:	Main Epistemology, metaphysics, ethics, logic
Notable ideas:	Categorical imperative, transcendental idealism, synthetic *a priori*, Noumenon, Sapere aude, nebular hypothesis

Carl Friedrich Gauss

Carl Friedrich Gauss (1777–1855), painted by Christian Albrecht Jensen

Born:	April 30, 1777 Braunschweig, Electorate of Brunswick-Lüneburg, Holy Roman Empire
Died:	February 23, 1855 (aged 77) Göttingen, Kingdom of Hanover
Residence:	Kingdom of Hanover
Nationality:	German
Fields:	Mathematician and physicist
Institutions:	University of Göttingen
Alma mater:	University of Helmstedt
Doctoral advisor:	Johann Friedrich Pfaff
Other academic advisors:	Johann Christian, Martin Bartels
Doctoral students:	Friedrich Bessel, Christoph Gudermann, Christian Ludwig Gerling, Richard Dedekind, Johann Encke, Johann Listing Bernhard Riemann, Christian Peters, Moritz Cantor
Other notable students:	Ferdinand Eisenstein, Gustav Kirchhoff, Ernst Kummer, Johann Dirichlet, August Ferdinand Möbius, Julius Weisbach, L.C. Schnürlein
Influenced:	Sophie Germain
Notable awards:	Copley Medal (1838)

Albert Einstein

Born:	March 14, 1879 Ulm, Kingdom of Württemberg, German Empire
Died:	April 18, 1955 (aged 76) Princeton, New Jersey, United States
Residence:	Germany, Italy, Switzerland, United States
Ethnicity:	Jewish
Citizenship:	Württemberg/Germany (until 1896)
	Stateless (1896–1901)
	Switzerland (from 1901)
	Austria (1911–1912)
	Germany (1914–1933)
	United States (from 1940)
Alma mater:	ETH Zurich University of Zurich
Known for:	General relativity and special relativity
	Photoelectric effect
	Mass-energy equivalence Quantification of the Brownian motion Einstein field equations
	Bose–Einstein statistics Unified Field Theory
Spouse:	Mileva Marić (1903–1919)
	Elsa Löwenthal, née Einstein (1919–1936)
Awards:	Nobel Prize in Physics (1921)
	Copley Medal (1925)
	Max Planck Medal (1929)
	Time Person of the Century

Sir Arthur C. Clarke, CBE

Arthur C. Clarke at his home office in Colombo, Sri Lanka, March 28, 2005

Born:	December 16, 1917 Minehead, Somerset, England, United Kingdom
Died:	March 19, 2008 (aged 90) Colombo, Sri Lanka
Pen name:	Charles Willis E.G. O'Brien
Occupation:	Author, inventor
Nationality:	British
Citizenship:	United Kingdom and Sri Lanka
Alma mater:	King's College London
Genres:	Hard science fiction, Popular science
Subjects:	Science
Notable work(s):	*Childhood's End 2001: A Space Odyssey, Rendezvous with Rama, The Fountains of Paradise*
Spouse(s):	Marilyn Mayfield (1953–1964)
Influences:	H.G. Wells, Jules Verne, Lord Dunsany, Olaf Stapledon
Influenced:	Stephen Baxter
	clarkefoundation.org

Index